Health and Safety Executive

Asbestos: The survey guide

London: TSO

Published by TSO (The Stationery Office), part of Williams Lea, and available from:

Online
www.tsoshop.co.uk

Mail, Telephone & E-mail
TSO
PO Box 29, Norwich, NR3 1GN
Telephone orders/General enquiries: 0333 202 5070
E-mail: customer.services@tso.co.uk
Textphone: 0333 202 5077

Published for the Health and Safety Executive under licence from the Controller of His Majesty's Stationery Office.

© Crown copyright 2012

First published 2010
Second Edition 2012

ISBN 978 0 7176 6502 0

This information is licensed under the Open Government Licence v3.0. To view this licence, visit http://www.nationalarchives.gov.uk/doc/open-government-licence/**OGL**

Any enquiries regarding this publication should be sent to: HSE.Online@hse.gov.uk

Some images and illustrations in this publication may not be owned by the Crown and cannot be reproduced without permission of the copyright owner. Where we have identified any third party copyright information you will need to obtain permission from the copyright holders concerned. Enquiries should be sent to HSE.Online@hse.gov.uk

Printed in the United Kingdom for The Stationery Office.
SD000332 8/25

This guidance is issued by the Health and Safety Executive. Following the guidance is not compulsory, unless specifically stated, and you are free to take other action. But if you do follow the guidance you will normally be doing enough to comply with the law. Health and safety inspectors seek to secure compliance with the law and may refer to this guidance.

Contents

How to use this guidance	iv
1 Introduction	1
Legal requirements	2
Appointed person	3
Health and safety issues	4
2 Competence and quality assurance procedures	5
3 Asbestos surveys	9
Purpose	9
Presumption or identification of ACMs	9
Types of survey	10
Survey restrictions and caveats	13
Survey strategy	13
4 Survey planning	16
Dutyholder's planning	16
Surveyor's planning procedure	16
Step 1: Collect all the relevant information to plan the survey	17
Step 2: Consider the information (desk-top study)	18
Step 3: Prepare a survey plan (including how data will be recorded)	18
Step 4: Conduct a risk assessment for the survey	19
5 Carrying out the survey (surveying)	21
Introduction	21
Bulk sampling strategy	23
Bulk sampling procedures	25
Bulk sampling	25
Material assessment	27
6 Survey report	28
Executive summary	28
Introduction	28
General site information	28
Survey results	28
Conclusions and actions	30
Bulk analysis results	30
7 Dutyholder's use of survey information	31
Appendix 1: Refurbishment and demolition surveys	32
Appendix 2: ACMs in buildings listed in order of ease of fibre release	35
Appendix 3: What ACMs look like and where to find them	41
Appendix 4: Material assessment algorithm	51
Appendix 5: Example of a survey and sampling equipment checklist	52
Appendix 6: Quality assurance and quality control	53
References	55
Further information	56

How to use this guidance

Green summary boxes: This publication has specific guidance for clients/dutyholders in green boxes:

Box 1: The purpose of an asbestos survey.

Box 3: What the client/dutyholder should do to check the competency of the surveyor.

Box 4: Areas to be inspected as part of a management survey.

Box 6: Information the client/dutyholder should expect from the surveyor.

Box 9: Information required for a management survey.

Box 10: Information required for a refurbishment or demolition survey.

Box 11: What the client/dutyholder should do to check the accuracy of the survey report.

Blue summary boxes: This publication has specific guidance for surveyors in blue boxes

Box 2: Survey key points.

Box 5: Information the surveyor needs from the client.

Box 7: Information to be collected by the surveyor.

Box 8: Example of a systematic survey inspection.

Box 1: The purpose of an asbestos survey

- To help manage asbestos in your premises.
- To provide accurate information on the location, amount and condition of asbestos-containing materials (ACMs).
- To assess the level of damage or deterioration in the ACMs and whether remedial action is required.
- To use the survey information to prepare a record of the location of any asbestos, commonly called an asbestos register,* and an asbestos plan of the building(s).
- To help identify all the ACMs to be removed before refurbishment work or demolition.

*Note: the information in the register should be used to inform the risk assessment (eg consider who could disturb asbestos on your premises), and to establish the management plan to prevent such a disturbance.

Box 2: Survey key points

- Be aware that the survey is essential for the client/dutyholder to successfully manage asbestos.
- All asbestos should be located as far as reasonably practicable within the survey type.
- Ensure that the appropriate survey is undertaken for the client's needs.
- Avoid caveats.
- Ensure the survey is reported in a format that can be used to prepare an asbestos register and building plan.
- Inform the client that the survey is not the end point in managing asbestos.

1 Introduction

1 This guidance has been prepared by the Health and Safety Executive (with the help of others, see Acknowledgements) to help people carrying out asbestos surveys and those with specific responsibilities for managing the risks from asbestos in non-domestic premises under regulation 4 of the Control of Asbestos Regulations 2012 (CAR 2012).[1] It is also designed to provide guidance in situations where surveys may be carried out for other purposes, eg for 'managing' asbestos in domestic premises under wider health and safety legislation and for meeting the requirements of the Construction (Design and Management) Regulations 2007 (CDM).[2] It complements and supports other guidance on managing asbestos.[3-5]

2 Large amounts of asbestos-containing materials (ACMs) were used for a wide range of construction purposes in new and refurbished buildings until 1999 when all use of asbestos was banned. This extensive use means that there are still many buildings in Great Britain which contain asbestos. Where asbestos materials are in good condition and unlikely to be disturbed they do not present a risk. However, where the materials are in poor condition or are disturbed or damaged, asbestos fibres are released into the air, which, if breathed in, can cause serious lung diseases, including cancers.

3 Workers who disturb the fabric of buildings during maintenance, refurbishment, repair, installation and related activities may be exposed to asbestos every time they unknowingly work on ACMs or carry out work without taking the correct precautions. The purpose of managing asbestos in buildings is to prevent or, where this is not reasonably practicable, minimise exposure for these groups of workers and other people in the premises. To prevent this exposure, information is needed on whether asbestos is, or is likely to be, present in the buildings, so that an assessment can be made about the risk it presents and appropriate measures put in place to manage those risks.

4 This guidance is aimed at:

- **Surveyors who carry out asbestos surveys.** It sets out how to survey premises for ACMs. In particular, it specifies the methodology to use in carrying out surveys and how to report and present the results. It also gives advice on how to recognise and sample suspected ACMs. In doing so, the guidance builds on and updates MDHS100 *Surveying, sampling and assessment of asbestos-containing materials*, which it replaces. It also contains a specific section which outlines the survey strategy to use when surveying large numbers of similar properties (eg domestic housing).
- **Those who commission surveys (eg clients/ dutyholders).** It sets out how to decide what type of survey is appropriate, how to select a competent surveyor, what the client should expect from a surveyor and what the client should provide to the surveyor. It also highlights issues (eg restricted access, excluded areas and other caveats) which not only reduce the effectiveness of the survey, but also have serious implications for managing asbestos. It also explains what checks should be made on the survey report to ensure its validity and accuracy (ie 'contract management').

5 The guidance will also be useful to building professionals, such as architects, designers, building surveyors and particularly demolition and asbestos removal contractors. For example, architects and building surveyors need to be aware of the requirement to carry out asbestos buildings surveys (and indeed can advise on the need for an asbestos survey before refurbishment and demolition projects). They should also be aware of the various types of surveys and be able to review completed surveys. Contractors need to be able to interpret asbestos surveys so that refurbishment or demolition can be planned and carried out safely.

6 The guidance does not cover airborne sampling or surveying contaminated land. These are specialised subjects outside the scope of this document.

Legal requirements

The duty to manage asbestos in non-domestic premises*

7 Asbestos, a category 1 human carcinogen, is subject to two sets of regulations – REACH (the Registration, Evaluation, Authorisation and Restriction of Chemicals Regulations 2007)[6] and, CAR 2012. REACH prohibits the importation, supply and use of asbestos. CAR 2012 covers work with asbestos, and licensing of asbestos-removal activities. Regulation 4 of CAR 2012 contains an explicit duty on the owners and occupiers of non-domestic premises, who have maintenance and repair responsibilities, to assess and manage the risks from the presence of asbestos (the duty is summarised in Figure 1). The risks will vary with circumstances and can arise from normal occupation of a building or from inadvertent disturbance during the repair, refurbishment and demolition of premises. The risk assessment will be used to produce a management plan which details and records what actions to take to manage and reduce the risks from asbestos.

8 The requirements are placed on 'dutyholders', who should:

- take reasonable steps to determine the location of materials likely to contain asbestos;
- presume materials to contain asbestos, unless there are good reasons not to do so;
- make and maintain a written record of the location of the ACMs and presumed ACMs;
- assess and monitor the condition of ACMs and presumed ACMs;
- assess the risk of exposure from ACMs and presumed ACMs and prepare a written plan of the actions and measures necessary to manage the risk (ie the 'management plan'); and
- take steps to see that these actions are carried out.

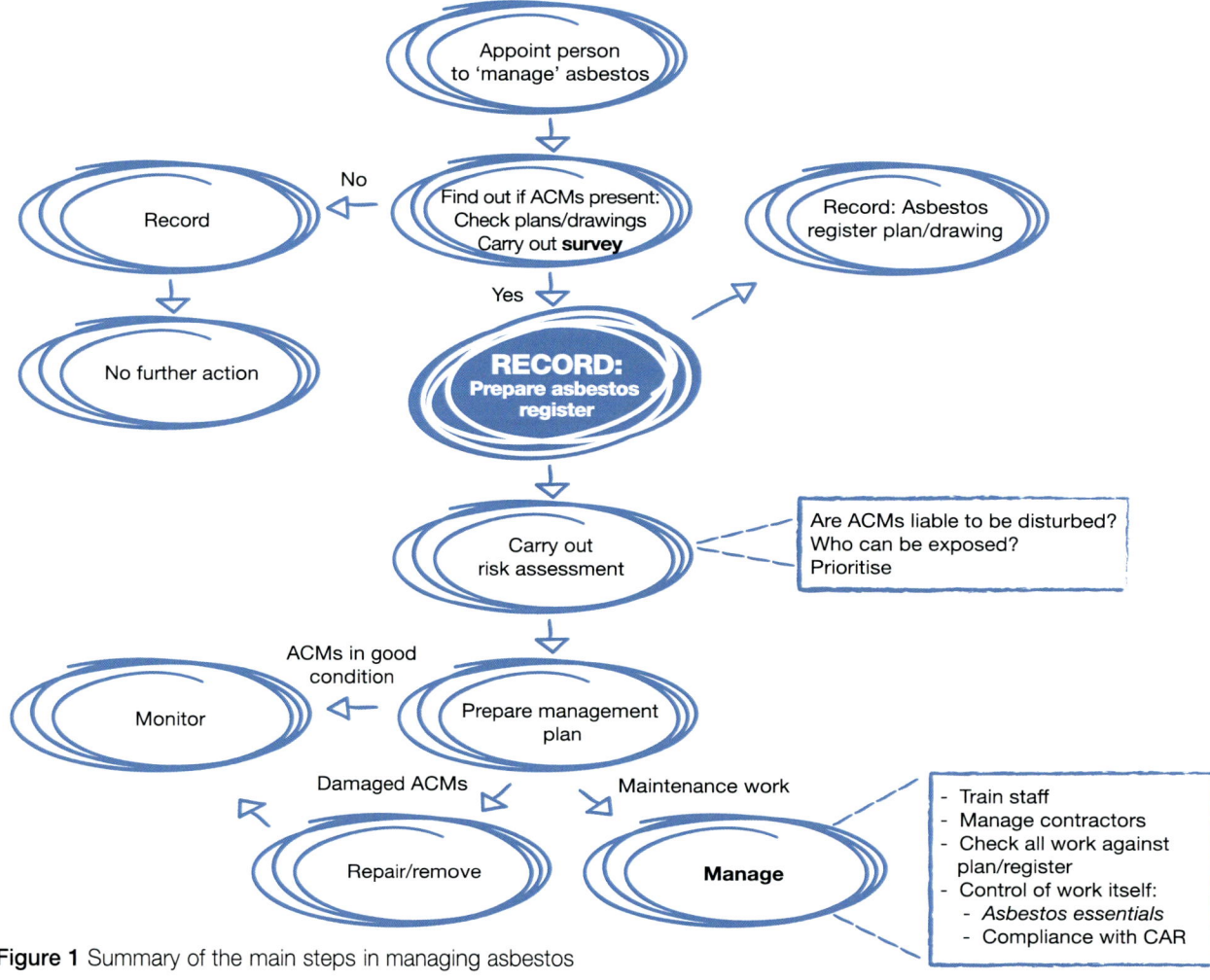

Figure 1 Summary of the main steps in managing asbestos

*The term 'premises' has a specific definition under health and safety legislation and includes vehicles, vessels, aircraft, installations on land and offshore, tents and moveable structures. While in most cases the survey will only be needed on existing buildings (including basements, cellars, tunnels, undercrofts etc) and the surrounding site, there may be some situations where there are hidden underground structures or pipes which may only come to light when refurbishment or demolition work is to take place. These should be included in the survey as appropriate.

9 To manage the risk from ACMs, the dutyholder will need to:

- keep and maintain an up-to-date record of the location, condition, maintenance and removal of all ACMs on the premises;
- repair, seal or remove ACMs if there is a risk of exposure due to their condition or location;
- maintain ACMs in a good state of repair and regularly monitor their condition;
- inform anyone who is liable to disturb the ACMs about their location and condition;
- have arrangements and procedures in place so that work which may disturb the ACMs complies with CAR 2012; and
- review the plan at regular intervals and make changes if circumstances change.

Management of asbestos in domestic premises

10 The 'duty to manage asbestos' requirements of regulation 4 of CAR 2012 do not normally apply to domestic premises. However, the requirements do apply to common parts of premises, including housing developments and blocks of flats, but do not place any direct duties on landlords for individual houses or flats. Examples of common parts would include foyers, corridors, lifts and lift shafts, staircases, boilerhouses, vertical risers, gardens, yards and outhouses. The requirements do not apply to rooms within a private residence which are shared by more than one household, such as bathrooms, kitchens etc in shared houses and communal dining rooms and lounges in sheltered accommodation.

11 The Health and Safety at Work etc Act 1974[7] section 2, requires all employers to conduct their work so their employees will not be exposed to health and safety risks, and to provide information to other people about their workplace which might affect their health and safety. Section 3 places duties on employers and the self-employed towards people not in their employment and section 4 contains general duties for anyone who has control, to any extent, over a workplace. In addition, the Management of Health and Safety at Work Regulations 1999[8] require employers to assess the health and safety risks to third parties, such as tenants who may be affected by their activities, and to make appropriate arrangements to protect them.

12 These requirements mean that organisations such as local authorities, housing associations, social housing management companies and others who own, or are responsible for, domestic properties, have legal duties to ensure the health and safety of their staff (and others) in domestic premises used as a place of work. As employers, the organisations also have duties under the general requirements of CAR 2012 to identify asbestos, carry out a risk assessment of work liable to expose employees to asbestos and prepare a suitable written plan of work.

Construction work

13 CDM requires arrangements to be in place to deal with asbestos during construction work, including refurbishment and demolition. Where construction or building work is to be carried out, the CDM client must provide designers and contractors who are bidding for the work (or who they intend to engage) with project-specific information about the presence of asbestos, so that the risks associated with design and construction work, including demolition, can be addressed. It is not acceptable to make general reference to hazards that may exist. Therefore, site-specific asbestos surveys should be carried out in advance of construction work to make sure that the information is available to those who need it.

Appointed person

14 To help comply with the legal requirements and to ensure that ACMs in premises are properly managed, dutyholders should identify a person (and in some cases a deputy) within their organisation who will be responsible for that management. An appointed person will be essential where the dutyholder has a large or complex building portfolio. The appointed person will need the resources, skills, training and authority to ensure that the ACMs are managed effectively. Part of their responsibilities will include managing the survey, including contractual and reporting arrangements, quality and subsequent use of the data.

15 The survey data and information will be used to complete an asbestos register and building diagram(s) showing the ACM locations. It will also feed into the risk assessment, which will be used to develop the management plan. The dutyholder needs to establish clear lines of responsibility for asbestos management and implementation of the plan.

Health and safety issues

16 Surveying and sampling ACMs can give rise to exposure to asbestos. These work activities are covered by the more general requirements of CAR 2012. The regulations require employers to carry out a risk assessment (regulation 6) and prepare a plan of work (regulation 7), setting out the control measures and personal protective equipment (PPE) to be used. The regulations also require that adequate information, instruction and training (including refresher training) (regulation 10) are given to the sampling personnel. Training should meet the requirements for non-licensable asbestos work as set out in the Approved Code of Practice, *Work with materials containing asbestos*. Sampling ACMs is, however, exempt from the regulations covering licensing (regulation 8), notification of work with asbestos (regulation 9) and health surveillance (regulation 22) by virtue of regulation 3(2), as the exposure is sporadic and low intensity and is unlikely to exceed the control limit. Other hazards may also be present, such as working at heights and electrical cables. A risk assessment will need to be carried out before starting work on site (see paragraphs 83–87). It should include any safety aspects and record any safety protocol to be observed on site as well as fire alarm and evacuation procedures.

> Asbestos surveying and sampling is likely to be 'work' with asbestos and therefore will require a risk assessment and a plan of work (method statement) under CAR 2012. Some activities may also involve physical work with asbestos (eg moving asbestos insulating board (AIB) ceiling tiles) and will require similar consideration.
>
> Some direct work on asbestos to support the survey may have to be carried out by a licensed asbestos contractor (see Appendix 1, paragraphs 6 and 7).

2 Competence and quality assurance procedures

17 Surveys can be carried out by in-house personnel or a third party. In each case, the surveyor must be competent to carry out the work required. To be competent, the 'surveyor' must:

- have sufficient training, qualifications, knowledge, experience and ability to carry out their duties in relation to the survey and to recognise their limitations;
- have sufficient knowledge of the specific tasks to be undertaken and the risks which the work will entail;
- be able to demonstrate independence, impartiality and integrity;
- have an adequate quality management system; and
- carry out the survey in accordance with recommended guidance (ie this publication).

> **HSE strongly recommends the use of accredited or certificated surveyors for asbestos surveys.**
>
> **The dutyholder should not appoint or instruct an independent surveyor to carry out a survey unless the surveyor is competent.**

18 Surveyors should have training and experience in all aspects of survey work including survey planning, resources, technical specification, quality control and ACM assessment criteria.

19 The asbestos surveyor needs knowledge of asbestos products (eg their nature, uses, hazards, sampling techniques etc) and also knowledge of building construction, construction methods, fire protection and the various uses of buildings. Surveyors should be aware of the different forms of building construction (eg system build, traditional, industrial etc) and how construction techniques affect asbestos use. Surveyors should also have knowledge of the use of ACMs in fire protection systems and the effect of building services on the distribution and location of ACMs. For example:

- fire protection in steel-framed buildings around columns and beams;
- fire protection around electrical and heating systems;
- fire protection separating multi-occupancy buildings;
- fire protection in lift shafts and risers;
- building services in voids, plenums, ducts, cavities, undercrofts and risers.

20 Surveyors should be aware of the range of building components and structures which contain asbestos (eg barge boards, chimney cowls, ducts, eaves, fascias, fire dampers, flue terminals and risers, gables, plenums, soffits, stud partitions, sandwich partitions etc).

21 Knowledge of building construction techniques and design is particularly relevant for refurbishment and demolition surveys, to understand where (and why) ACMs may have been used in a structure. Surveyors should also be aware that there are many unrecorded ad hoc uses of ACMs in buildings. Some uses arose simply from the convenient presence of ACMs as building and engineering materials. AIB panels and offcuts, for example, were used extensively, randomly and imaginatively as shuttering for concrete, packers around columns, spacers around window and door frames, and cavity closers. Other ACMs may have caused contamination in buildings from the way they were applied, poor work practices or later disturbance, producing for example:

- overspray and spread of dust from sprayed coatings;
- residues from thermal insulation on brickwork and in ducts;
- debris from AIB fire breaks in ceiling voids and also in cavity walls.

These ACMs are often hidden and unrecorded in building plans.

22 Survey thoroughness is important. Simple and obvious ACMs are sometimes missed, as well as those which are hidden or obscured. The survey should be performed in a structured, methodical and systematic manner. The use of checklists and a structured approach to the survey process will minimise the risk of ACMs being missed (see Box 8). Adequate time must be allowed for the survey inspection to be done effectively.

> The dutyholder must ensure that adequate time and resources are made available to the surveyor(s) to allow a thorough survey to be carried out.

23 Organisations can demonstrate that they are technically competent to undertake surveys for ACMs through accreditation to ISO/IEC 17020.[9] The United Kingdom Accreditation Service (UKAS) is the sole national accreditation body in the United Kingdom (UKAS, 21–47 High Street, Feltham, Middlesex TW13 4UN Tel: 020 8917 8400 www.ukas.com). Accreditation gives an assurance that an independent and authoritative body has assessed the technical competence of an organisation, including its underpinning management system. The scheme should ensure that the organisation can provide a valid service for the services specified on its schedule of accreditation.

24 Individual surveyors can also demonstrate that they are technically competent to undertake specified surveys through holding 'personnel' certification from a Certification Body accredited by UKAS for this activity under ISO/IEC 17024.[10] Personnel certification provides assurance that an individual has achieved a defined level of competence to carry out specific activities. Currently there is no accredited scheme in operation. A number of people may have been certificated under previous schemes, 'NIACS' (National Individual Asbestos Certification Scheme) and 'ABICS' (Asbestos Building Inspectors Certification Scheme). Certificated surveyors should also work within a general Quality Assurance framework provided by ISO/IEC 17020 (or ISO 9001[11] as a minimum).

> Accreditation and personnel certification are both valid schemes for demonstrating competence in performing asbestos surveys. The schemes are designed for different market segments and have different emphasis. Both will ensure that surveys are carried out by competent people. Accreditation is suitable for organisations of all sizes where the scale and volume of surveying work dictates not only individual competence but also the need for more formal and well-defined quality management systems.
>
> Personnel certification is designed for individuals who may operate as sole traders or in organisations with only a few surveyors. The scheme focuses on individual competence.

25 Individuals without personnel certification may be able to demonstrate that they have sufficient competency to undertake specified surveys through a combination of qualifications and experience. In this situation, experience (ie extent and range) is particularly important. The most widely held training qualification in the UK is the BOHS Proficiency Module P402: 'Buildings surveys and bulk sampling for asbestos' (other proficiency courses may be available from other training organisations). The P402 is a basic minimum qualification for individuals carrying out asbestos surveys and on its own it does not demonstrate competency. Therefore, in addition, individuals must also have at least six months' full-time, relevant, practical field experience on asbestos surveys under the supervision of experienced and suitably qualified personnel. The experience should cover the property sectors including industrial, commercial and domestic, and should cover management surveys and refurbishment and demolition surveys, as appropriate. Trainees will be able to demonstrate a certain level

of competence through audit or assessment on an appropriate number of surveys (eg at least five) before they can be allowed to operate as the lead surveyor.

26 Further training and experience will be necessary to ensure competence in refurbishment and demolition surveys particularly for large premises. Training should cover, for example, the potential additional locations to be inspected, access techniques into cavities, walls and partitions, sandwich partitions etc. The Proficiency Module P402 can be supplemented with two refresher modules, P402R, relating to management surveys, and refurbishment and demolition surveys respectively. These modules can be a useful way of providing ongoing annual refresher training, as well as the opportunity to exchange information and experience with others (as required in ISO/IEC 17020).

The P402 qualification on its own does not demonstrate competency. Individuals must have at least six months' full-time, relevant, practical field experience on asbestos surveys under the supervision of experienced and suitably qualified personnel.

27 The BOHS S301 course ('Asbestos and other fibres') is also a relevant starting qualification, but again on its own does not demonstrate an individual's competence. However, individuals can then obtain a Certificate of Competence in Asbestos (CoCA) from BOHS after obtaining the S301, by completing six months' practical experience in asbestos, successfully submitting a written report (eg on asbestos surveys) and passing an oral exam. The qualification must still be supplemented by adequate supervised field experience.

28 Personnel may also hold another qualification in surveying: the Royal Society for Public Health (RSPH) Level 3 Certificate in asbestos inspection procedures. This qualification was developed as part of a personnel certification scheme and is still available on the RSPH website (www.rsph.org.uk). The qualification alone does not demonstrate competency. Candidates will also need at least six months' supervised and audited practical experience, as outlined for the P402 qualification (see paragraph 25).

29 All surveying organisations should have a quality management system (ie quality assurance and quality control schemes) in place to ensure the highest standards. These schemes should be written and should include a minimum of these three component parts:

- A proportion of surveys being reinspected by another competent surveyor/auditor, usually while the survey is in progress. All aspects of the site work (safety assessments, inspection procedures, sampling, documentation, material risk assessments etc) should be checked. It is recommended that about 5% of surveys are reinspected (BS 6002: 2006).[12]
- All the management procedures and systems of a surveying organisation should be quality assured by carrying out audits of completed surveys. This would normally be a desk-top audit.
- There should also be a quality control scheme for survey reports. All reports should be checked before being issued to clients. Simple but thorough checks should be made that the client's requirements have been met, as well as checks on the consistency, technical accuracy and completeness of the report.

30 More details of a quality management system are given in Appendix 6.

31 Laboratories who carry out bulk analysis for asbestos must demonstrate that they conform to the requirements of ISO/IEC 17025[13] and, if they provide this service for a third party, must be accredited by a recognised accreditation body, ie UKAS. The laboratory should be able to demonstrate its competence to carry out bulk asbestos analysis through:

- staff training records;
- certificates from external training providers;
- participation in quality assurance schemes;
- internal proficiency testing programmes;
- satisfactory performance in national proficiency testing programmes;
- replicate analysis checks of a proportion of the routine samples.

32 The Asbestos in Materials Scheme (AIMS) is the UK national proficiency testing programme for bulk asbestos analysis. Individual analysts should also demonstrate competency through training records and satisfactory performance in an internal quality assurance scheme.

33 It is the responsibility of anyone using a laboratory for the analysis of samples for asbestos to make sure the lab holds the necessary accreditation (details can be obtained from the UKAS website: www.ukas.com).

34 Samples or representative sub-samples should be kept for at least six months after analysis to allow checks to be made. Samples associated with a legal dispute or claim may need to be kept for longer.

Box 3: What the client/dutyholder should do to check the competency of the surveyor

The dutyholder should be satisfied that the surveyor is competent to carry out the work required.

This means that the dutyholder should make reasonable enquiries as to whether the organisation or individual is technically competent to carry out the survey adequately and safely, and can allocate adequate resources to it. The competency enquiry should be carried out as a two-stage process:

- **Stage 1:** An assessment of the individual's or company's survey expertise and also, their knowledge of health and safety, to determine whether these are sufficient to enable them to carry out the survey competently, safely and without risk to health.
- **Stage 2:** An assessment of the individual's or company's experience and track record to establish if it is capable of doing the work and that it recognises its limitations.

Stage 1: Establish the accreditation or certification status of the surveyor and any relevant asbestos survey qualifications (see paragraphs 23–28). Obtain a written declaration which states that the surveyor can operate with independence, impartiality and integrity and that personnel carrying out the work are adequately trained for all aspects of the work taking place. In addition, obtain copies of the current insurance certificates for employer's liability, public liability and professional indemnity cover and check them to see that they cover the proposed work.

Stage 2: Obtain information on the surveyors' past experience on the type of survey planned and their capability to do the work. References or evidence of recent similar work should be requested.

If a company or surveyor cannot demonstrate competence through current accreditation or personnel certification, the dutyholder will need to conduct a more detailed assessment of their competence to do the work. This will include requesting: details of their qualifications, copies of their written procedures (including quality control policies) and references to other evidence of recent similar work.

3 Asbestos surveys

Purpose

35 The purpose of the survey is to help manage asbestos in the dutyholder's premises. The survey has to provide sufficient information for: an asbestos register and plan to be prepared, a suitable risk assessment to be carried out and a written plan to manage the risks to be produced. The process is shown schematically in Figure 2.

36 In most cases, the survey will have three main aims:

- it must as far as reasonably practicable locate and record the location, extent and product type of any presumed or known ACMs;
- it must inspect and record information on the accessibility, condition* and surface treatment of any presumed or known ACMs;
- it should determine and record the asbestos type, either by collecting representative samples of suspect materials for laboratory identification, or by making a presumption based on the product type and its appearance etc.

Presumption or identification of ACMs

37 The duty to manage requirement in CAR 2012 regulation 4 allows materials to be 'presumed' to contain asbestos. Therefore in the asbestos survey, materials can be presumed to contain asbestos. There are two levels of 'presumption':

1 Strong presumption: in this case the material looks as if it is an ACM, or that it might contain asbestos. This conclusion can be reached through visual inspection alone by an experienced, well-trained surveyor, familiar with the range of asbestos products. Examples of 'strong presumption' would be:

- where laboratory analysis has confirmed the presence of asbestos in a similar construction material;
- materials in which asbestos is known to have been commonly used in the manufactured product at the time of installation (eg corrugated cement roof and wall sheeting, cement gutters and drainpipes, cement water tanks, ceiling tiles, insulating boards);
- materials which have the appearance of asbestos but no sample has been taken, eg thermal insulation on a pipe where fibres are clearly visible.

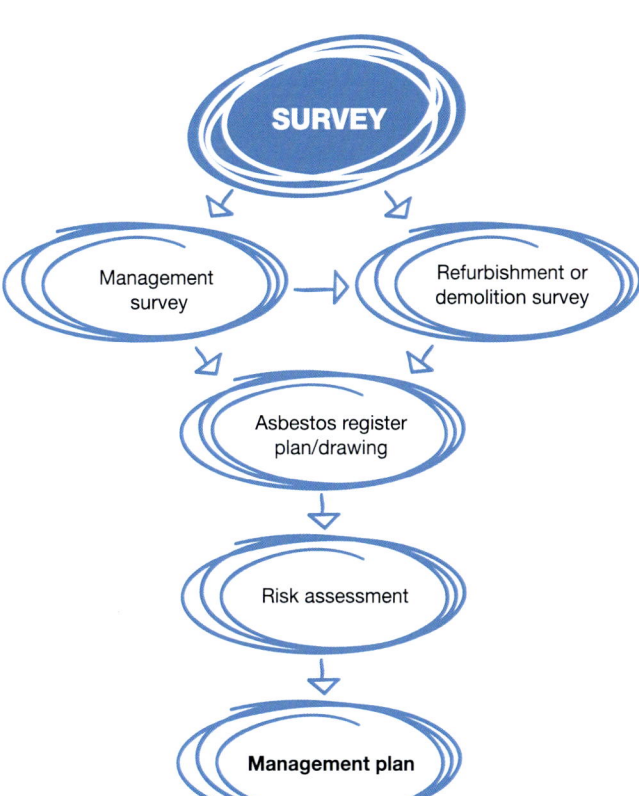

Figure 2 Schematic diagram showing the link between the asbestos survey and the management plan

---

* The exception is refurbishment and demolition surveys where information on the condition of the asbestos is usually **not** required (see paragraph 52), as the ACM will be removed soon after the survey. However, in circumstances where the removal will not take place for some time after the survey (eg more than three months), the ACMs will have to be managed during this period. In this situation, the condition of the ACMs should also be determined and remedial action taken as appropriate (see paragraphs 124 and 130).

2 A **'default' situation** where a material is **presumed** to contain asbestos because there is insufficient evidence (eg no analysis) to confirm that it is asbestos free, or where a dutyholder/surveyor decides that it is easier under the planned management arrangements to presume certain materials contain asbestos. Many non-asbestos materials will also be presumed to contain asbestos using this system. There is a further default situation where materials must be presumed to contain asbestos. The default applies to areas which cannot be accessed or inspected. In this situation **any area not accessed or inspected must be presumed to contain asbestos, unless there is strong evidence that it does not.**

38 Materials cannot be presumed to be asbestos free (ie contain no asbestos) unless there is strong evidence to conclude that they are highly unlikely to contain asbestos. There are obvious materials which are not asbestos, eg wood, glass, metal, stone etc. There are also many examples of asbestos being present inside materials, eg a sandwich layer inside doors, inside columns or under column casings, on the 'hidden' side of items, eg wood panelling, ceiling tiles, under veneers. Reasons to conclude that a material does not contain asbestos would be:

- non-asbestos substitute materials were specified in the original architect's/quantity surveyor's plans or in subsequent refurbishments;
- the product was very unlikely to contain asbestos or have asbestos added (eg wallpaper, plasterboard etc);
- post-1985 construction (for amphibole ACMs such as asbestos insulating board, see Appendix 1);
- post-1990 construction for decorative textured coatings (formulations containing asbestos were prohibited in 1988 and some suppliers voluntarily ceased using asbestos in 1984);
- post-1999 construction (some chrysotile products were prohibited in 1993 and nearly all were prohibited in 1999).

39 It is not always straightforward to conclude that ACMs are absent. The regulations require that reasonable steps are taken. While original specifications may not have included ACMs in certain building locations, workers may have used them for their convenience. For example, work on building systems (eg CLASP systems[14] has shown that ACMs, eg asbestos insulating board (AIB) off-cuts were used as filler/packing and support items in places where their presence was not recorded. There are also many examples of poor removal practice leaving asbestos-containing debris and residues. Therefore areas where asbestos has been removed previously will need to be reinspected as part of the survey.

> Areas where asbestos has been removed previously will need to be reinspected.

Types of survey

40 This document describes two different types of survey: **management surveys** and **refurbishment and demolition surveys**.

41 The type of survey will vary during the lifespan of the premises and several may be needed over time. A management survey will be required during the normal occupation and use of the building to ensure continued management of the ACMs in situ. A refurbishment or demolition survey will be necessary when the building (or part of it) is to be upgraded, refurbished or demolished. It is probable that at larger premises a mixture of survey types will be appropriate, eg a boiler house due for demolition will require a refurbishment/demolition survey, while offices at the same site would have a management survey. In later years refurbishment surveys may be required in rooms or floors which are being upgraded. In sectors where there are large numbers of properties (eg domestic houses) or internal units (eg hotels), only particular rooms may be specified for upgrading, eg kitchens, bathrooms and bedrooms. Refurbishment surveys would only be necessary in these locations.

42 It is important that the client and the surveyor know exactly what type of survey is to be carried out and where, and what the specification will be. So there should be a clear statement and record of the type of survey that is to be carried out, including the reasons for selecting that type of survey, and where it is to be carried out.

Management survey

43 A management survey is the standard survey. Its purpose is to locate, as far as reasonably practicable, the presence and extent of any suspect ACMs in the building which could be damaged or disturbed during normal occupancy, including foreseeable maintenance and installation, and to assess their condition.

44 Management surveys will often involve minor intrusive work and some disturbance. The extent of intrusion will vary between premises and depend on what is reasonably practicable for individual properties, ie it will depend on factors such as the type of building, the nature of construction, accessibility etc. A management survey should include an assessment of the condition of the various ACMs and their ability to release fibres into the air if they are disturbed in some way. This 'material assessment' (see paragraphs 124–127) will give a

good initial guide to the priority for managing ACMs as it will identify the materials which will most readily release airborne fibres if they are disturbed.

45 The survey will usually involve sampling and analysis to confirm the presence or absence of ACMs. However, a management survey can also involve presuming the presence or absence of asbestos. A management survey can be completed using a combination of sampling ACMs and presuming ACMs or, indeed, just presuming. Any materials presumed to contain asbestos must also have their condition assessed (ie a material assessment).

> **Management surveys can involve a combination of sampling to confirm asbestos is present or presuming asbestos to be present.**

46 By presuming the presence of asbestos, the need for sampling and analysis can be deferred until a later time (eg before any work is carried out). However, this approach has implications for the management arrangements. The dutyholder bears potential additional costs of management for some non-ACMs. Any work carried out on 'presumed' materials would need to involve appropriate contractors and work methods in compliance with CAR 2012 irrespective of whether the material was actually an ACM or not. Alternatively, before any work starts, sampling and analysis can be undertaken to confirm or refute the presence of asbestos. The results will determine the work methods and contractors to be used. The 'presumption' approach has several disadvantages: it is less rigorous, it can lead to constant obstructions and delays before work can start, and it is more difficult to control. 'Default' presumptions may also lead to unnecessary removal of non-ACMs and their disposal as asbestos waste. Default presumptions may be suitable in some instances, eg 'small' or simple premises, as part of a client's management arrangements.

47 Surveyors should always endeavour to positively identify ACMs. A sufficient number of samples should be taken to confirm the location and extent of ACMs. It is legitimate to reduce sample numbers where materials can be strongly presumed to be ACMs. However, the default presumption option should be avoided where possible, as it can make managing asbestos more difficult for the dutyholder. Default presumption should only be used in circumstances where it is requested by the client and/or where access genuinely cannot be obtained.

48 When sampling is carried out as part of a management survey, samples from each type of suspect ACM should be collected and analysed. If the material sampled is found to contain asbestos, other similar materials used in the same way in the building can be strongly presumed to contain asbestos. Less homogeneous materials (eg different surfaces/coating, evidence of repair etc) will require a greater number of samples. The sample number should be sufficient to establish whether asbestos is present or not in the particular material. Sampling may take place simultaneously with the survey, or as in the case of some larger surveys, can be carried out later as a separate exercise.

49 All areas should be accessed and inspected as far as is reasonably practicable. Areas should include underfloor coverings, above false ceilings, and inside risers, service ducts, lift shafts etc (see Box 4). **Surveying may also involve some minor intrusive work,** such as accessing behind fascia and panels and other surfaces or superficial materials. The extent of intrusion will depend on the degree of disturbance that is or will be necessary for foreseeable maintenance and related activities, including the installation of new equipment/cabling. Surveyors should come prepared to access such areas (ie with the correct equipment etc). Management surveys are only likely to involve the use of simple tools such as screwdrivers and chisels. Any areas not accessed must be presumed to contain asbestos. The areas not accessed and presumed to contain asbestos must be clearly stated in the survey report and will have to be managed on this basis (see paragraph 46), ie maintenance or other disturbance work should not be carried out in these areas until further checks are made.

> **Box 4: Areas to be inspected as part of a management survey**
>
> All ACMs should be identified as far as is reasonably practicable. The areas inspected should include: underfloor coverings, above false ceilings (ceiling voids), lofts, inside risers, service ducts and lift shafts, basements, cellars, underground rooms, undercrofts (this list is not exhaustive).

50 Management surveys should cover routine and simple maintenance work. However, it has to be recognised that where 'more extensive' maintenance or repair work is involved, there may not be sufficient information in the management survey and a localised refurbishment survey will be needed. A refurbishment survey will be required for all work which disturbs the fabric of the building in areas where the management survey has not been intrusive. The decision on the need for a refurbishment survey should be made by the dutyholder (probably with help from others).

Refurbishment surveys will be required for all work which disturbs the fabric of the building in areas where the management survey has not been intrusive.

The dutyholder will need to make the decision but probably with help from others.

Refurbishment and demolition surveys

51 A **refurbishment and demolition** survey is needed before any refurbishment or demolition work is carried out. This type of survey is used to locate and describe, as far as reasonably practicable, all ACMs in the area where the refurbishment work will take place or in the whole building if demolition is planned. The survey will be fully intrusive and involve destructive inspection, as necessary, to gain access to all areas, including those that may be difficult to reach. A refurbishment and demolition survey may also be required in other circumstances, eg when more intrusive maintenance and repair work will be carried out or for plant removal or dismantling.

52 There is a specific requirement in CAR 2012 (regulation 7) for all ACMs to be removed as far as reasonably practicable before major refurbishment or final demolition. Removing ACMs is also appropriate in other smaller refurbishment situations which involve structural or layout changes to buildings (eg removal of partitions, walls, units etc). Under CDM, the survey information should be used to help in the tendering process for removal of ACMs from the building before work starts. The survey report should be supplied by the client to designers and contractors who may be bidding for the work, so that the asbestos risks can be addressed. In this type of survey, where the asbestos is identified so that it can be removed (rather than to 'manage' it), the survey does not normally assess the condition of the asbestos, other than to indicate areas of damage or where additional asbestos debris may be present. However, where the asbestos removal may not take place for some time, the ACMs' condition will need to be assessed and the materials managed (see paragraph 124).

53 Refurbishment and demolition surveys are intended to locate all the asbestos in the building (or the relevant part), as far as reasonably practicable. It is a disruptive and fully intrusive survey which may need to penetrate all parts of the building structure. Aggressive inspection techniques will be needed to lift carpets and tiles, break through walls, ceilings, cladding and partitions, and open up floors. In these situations, controls should be put in place to prevent the spread of debris, which may include asbestos. Refurbishment and demolition surveys should only be conducted in unoccupied areas to minimise risks to the public or employees on the premises. Ideally, the building should not be in service and all furnishings removed. For minor refurbishment, this would only apply to the room involved or even part of the room where the work is small and the room large. In these situations, there should be effective isolation of the survey area (eg full floor to ceiling partition), and furnishings should be removed as far as possible or protected using sheeting. The 'surveyed' area must be shown to be fit for reoccupation before people move back in. This will require a thorough visual inspection and, if appropriate (eg where there has been significant destruction), reassurance air sampling with disturbance. Under no circumstances should staff remain in rooms or areas of buildings when intrusive sampling is performed.

54 There may be some circumstances where the building is still 'occupied' (ie in use) at the time a 'demolition' survey is carried out. For example in the educational sector, refurbishment/demolition surveys may be conducted in schools or colleges during one closure period (eg holidays) and the work not undertaken until the next holiday period. Also, a demolition survey maybe conducted to establish the economic future or viability of a building(s). The survey results would determine the outcome. In such situations, the 'survey' will need extremely careful managing with personnel and equipment/furnishings being decanted and protected (as necessary), while the survey progresses through the building. Again, there should be effective isolation of the survey areas and the 'surveyed' area must be shown to be fit for reoccupation before personnel reoccupy (see paragraph 53).

Survey restrictions and caveats

55 The value and usefulness of the survey can be seriously undermined where either the client or the surveyor imposes restrictions on the survey scope or on the techniques/method used by the surveyor. Information on the location of all ACMs, as far as reasonably practicable, is crucial to the risk assessment and development of the management plan. Any restrictions placed on the survey scope will reduce the extent to which ACMs are located and identified, incur delays and consequently make managing asbestos more complex, expensive and potentially less effective.

56 In management surveys, surveyors should be properly prepared for accessing all reasonably practicable areas in all parts of the building (see Box 4). Potentially difficult to enter areas (including locked rooms etc) should be identified in the planning stage with the dutyholder and arrangements made for access, eg mobile elevating work platforms (MEWPs) for work at height, rooms unlocked, doors/corridors unblocked etc. In situations where there is no entry on the day of the survey, a revisit should be made when access will be possible. Where there are health and safety risks associated with some activities (eg height, confined spaces), these should be adequately assessed and arrangements made to control them (see paragraphs 83–91). Any area not accessed (and where no other information exists) must be presumed to contain asbestos and be managed on that basis.

57 In refurbishment surveys, the area and scope of the work will need to be agreed between the dutyholder and the surveyor. In these surveys and in demolition surveys there should be no restrictions on access unless the site is unsafe (eg fire-damaged premises) or access is physically impractical. The level of intrusion will be significantly greater than with management surveys. It will include accessing structural areas, between floors and walls and underground services. Some areas may be difficult to gain entry to and/or may need specialist assistance or equipment. Access arrangements need to be fully discussed in the planning stage and form part of the contract, particularly where assistance has to be engaged. Where access has not been possible during refurbishment and demolition surveys, these areas must be clearly located on plans and in the text of the report to allow the refurbishment and demolition processes to be progressive in those areas. Any ACMs must be identified and removed at this time. It is now recognised that even with 'complete' access demolition surveys, all ACMs may not be identified and this only becomes apparent during demolition itself. Surveyors need to be competent to do all the relevant work and tasks in this class of surveys (see Section 2: Competence and quality assurance procedures). They will need some knowledge of construction, be able to carry out the work safely and without risk to health, have the correct equipment to do the work and have the appropriate insurance.

58 If any restrictions have to be imposed on the scope or extent of the survey, these items must be agreed by both parties and clearly documented. They should be agreed before work starts (eg at the preliminary site meeting and walk-through inspection or during discussion (see paragraphs 77–78)) and are likely to form part of the contract. If during the survey the surveyor is unable to access any location or area for any reason, the dutyholder must be informed as soon as possible and arrangements made for later access. If access is not possible, then the survey report should clearly identify these areas not accessed. Limitations should be kept to an absolute minimum by ensuring that staff are adequately trained, insured and have the appropriate equipment and tools.

> **Survey restrictions and caveats can seriously undermine the management of asbestos in buildings. They should be included only where absolutely necessary and should be fully justified. Most can be avoided by proper planning and discussion. They must be agreed between the dutyholder and the surveyor and documented in the survey report.**

Survey strategy

Non-domestic premises

59 In the non-domestic sector, there is an expectation that every building will be surveyed on an individual basis to identify the presence and condition of asbestos. In premises where there are large numbers of similar or near-identical rooms (eg offices or hotels), a survey strategy can be adopted which reflects the scale and nature of the buildings. All rooms should be visually inspected, as there clearly can be differences in rooms due to location (eg presence of risers, services) or function/facilities. Subsequently, 'similar' rooms can be placed into groups (ie rooms with similar locations or facilities, such as next to lifts, containing risers, gable end or middle building rooms, plant rooms etc). In these groups there is likely to be greater uniformity in the presence of ACMs, eg fire protection next to lift shafts). Within these groups, there will be less need for sampling in all rooms. Sampling can be conducted in a representative number of rooms and, where ACMs are identified, the same items in other rooms in this group can be strongly presumed to contain asbestos.

Domestic premises

60 In the domestic sector, local authorities and housing associations have responsibility for very large numbers of properties which need a range of maintenance and repair work as well as general improvement and upgrading or occasionally demolition. Works can include electrical rewiring, structural repairs and alterations, replacement windows, central heating, insulation, renewal of bathroom and kitchen fittings or complete renovations. The work may be necessary on individual or small numbers of premises (eg emergency work due to fire/water/storm damage) or on large numbers where there are major improvement or upgrading schemes.

61 Domestic properties present particular challenges for surveying asbestos. The main issues are the scale (ie large number of properties (and consequently what is reasonable and practicable)), the real extent of similarity in building materials and the personal nature of the property. Asbestos was extensively used in domestic properties between 1930 and 1980. However, the presence of ACMs can now be quite variable and unpredictable even within the same archetypal group. The content varies for several reasons including:

- inconsistent/variable initial use;
- random use of waste pieces and offcuts by builders;
- previous unrecorded removal of asbestos;
- modifications of properties by tenants (present and past) and housing associations (removing and adding ACMs).

62 Domestic dwellings often fall into particular archetypal groups in terms of style, design and age, eg flats within blocks would generally be similar at construction. These factors can be used to develop the survey strategy. The following paragraphs outline the general strategy to use for surveying domestic properties. There are three components: establish the asbestos status of properties, carry out management surveys and carry out, as necessary, refurbishment surveys.

Establish asbestos status of properties

63 Carry out a desk-top study to establish the probable asbestos status of groups of properties. In this exercise, properties can be placed into archetypal groups based on various parameters including construction date (eg estates phases), house design and location. These groups of properties can be separated into the following categories: asbestos free, 'contain' ACMs and 'possibly contain' ACMs. The main criteria involved here for concluding groups are asbestos free would be any property constructed in 2000 or later. It may also be possible to conclude that groups are asbestos free based on other information, such as original construction information, building material specification, previous asbestos surveys or removals or other records. The evidence for this would need to be strong and records complete. These sources of information would also be used to conclude the definite presence of asbestos in particular property groups. Other properties constructed pre-2000 should be classed as possibly containing ACMs (unless there is evidence to show otherwise (eg previous surveys etc)).

Management surveys

64 Management surveys should be carried out on properties which contain or possibly contain ACMs. 'Asbestos-free' dwellings should be recorded as such in the management plan and do not need surveying. However, workers in such premises (particularly pre-2000) should always be vigilant. A proportion of properties in each category (ie that contain or possibly contain ACMs) and each archetypal group should be surveyed. Exact sampling ratios cannot be specified, as the proportion will depend on the variability of housing stock. A proportion should be surveyed until the results demonstrate as far as reasonably practicable that there is consistency in the range of ACMs in the property type. Not every property will contain all the ACM items but the range of ACMs in the property types will be known. Every non-surveyed property has the potential to contain all the ACMs in the range and the ACMs should be managed on that basis. Where there is considerable variability, the ratio surveyed will be high.

65 Information from the management surveys can be enhanced with data from more intrusive surveys when the circumstances allow, eg when properties are vacant. Information from refurbishment and demolition surveys should be used to update the asbestos register for that particular type of property.

66 Management surveys, supported by refurbishment and demolition surveys, should be used as the primary means of managing routine maintenance work in domestic premises. However, dutyholders must recognise that these surveys are limited in their scope and extent of intrusion and therefore do not provide sufficient information on the presence of ACMs for larger scale refurbishment and other improvement projects.

Refurbishment surveys

67 Refurbishment and demolition surveys will be required where refurbishment work or other work involving disturbing the fabric of the building is carried out. The survey strategy for refurbishment works is similar to that for management surveys. Refurbishment and demolition surveys should also be carried out on a proportion of properties in the work programme. The ratio again will depend on asbestos variability within the housing stock and may be high where there is substantial variation. A proportion should be surveyed until the results demonstrate as far as reasonably practicable that there is consistency in the range of ACMs in the property type and there is an accurate picture of asbestos presence. The refurbishment and demolition survey will only be necessary in the specific area/location where the works will take place, eg cupboard, part of a room, kitchen/bathroom. However, further refurbishment and demolition surveys will be necessary in other locations when new improvement schemes are proposed. These localised refurbishment and demolition surveys should have the specific purpose of identifying ACMs for removal, control or avoidance during the refurbishment works.

68 For house improvement schemes and other project work, refurbishment and demolition surveys should be incorporated into the planning phase of such work as far as possible. This will avoid delays and disruption etc. Where the work is urgent (eg essential or emergency maintenance, repair and installation), the refurbishment surveys may have to be carried out just before the work itself. Surveys should be performed with due diligence.

69 The above strategy requires management arrangements which reflect the circumstances and uncertainty of ACMs in domestic premises. There will always be the potential for ACMs not to have been identified before maintenance and refurbishment work is carried out. In these situations the management arrangements must include the following:

- adequate asbestos training of tradespeople (eg to cover awareness, including identification) and work procedures;
- arrangements must be in place to ensure that asbestos registers or records are checked before work commencing and there are procedures for dealing with any suspect/suspicious/unknown material, ie stop work, check material etc;
- adequate supervision to ensure procedures are implemented and followed.

4 Survey planning

70 The key to an effective survey is the planning. The degree of planning and preparation will depend on the extent and complexity of the building portfolio. Single, simple one-storey factory buildings will be different from a school or a large hospital complex. Surveys on sites with many and variable types of buildings will need considerable planning and prioritising. The principles to be used in planning/structuring and conducting the survey will be similar in all cases. The survey is not about just turning up and taking samples. There needs to be a sufficient initial exchange of information between the dutyholder or client and the surveyor and a clear understanding by both parties of what is required. The information will be used to form the contract between the dutyholder and the surveyor including where the survey is performed in-house.

71 The key points are summarised in Boxes 5 and 6. The dutyholder should know what to expect from the surveyor and vice versa.

Dutyholder's planning

72 The dutyholder needs to consider the purpose of the survey and what information it needs to provide. The dutyholder will be the client and should consider:

- Why the survey is needed.
- What type(s) of survey is needed?
- What information must the survey provide?
- What format do I want the report in (asbestos register, drawings, electronic, printed etc)?
- What information will the surveyor require?

Surveyor's planning procedure

73 The surveyor should establish the type of survey(s) required. It may be that more than one survey type will be required, eg a management survey for most of the premises, but a refurbishment survey in one building or part of a building. Establishing the survey type should be done in consultation with the client. The survey planning should be structured and include the various steps outlined below. These steps are listed separately but in practice there will be overlap or they will run concurrently/simultaneously. There may be some situations where all the steps are not necessary or possible (eg small or simple premises, fire-damaged premises and pre-purchase surveys etc). Where the survey involves sampling or asbestos disturbance, a site-specific assessment and plan of work is required under CAR 2012.

Box 5: Information the surveyor needs from the client

- Details of buildings or parts of buildings to be surveyed and survey type(s).
- Details of building(s) use, processes, hazards, priority areas.
- Plans, documents, reports and surveys on design, structure and construction.
- Safety and security information: fire alarm testing, special clothing areas (eg food production).
- Access arrangements and permits.
- Contacts for operational or health and safety issues.

Box 6: Information the client/dutyholder should expect from the surveyor

- Surveyor(s) identity, qualifications, accreditation or certification status, quality control procedures.
- References from previous work.
- Insurance (professional indemnity cover).
- Costs.
- Proposed scope of work.
- Plan of work, including plans for sampling or asbestos disturbance.
- Timetable.
- Details of caveats.
- Report, including areas not accessed/not surveyed.

Step 1: Collect all the relevant information to plan the survey.

Step 2: Consider the information (desk-top study).

Step 3: Prepare a survey plan (including how data will be recorded).

Step 4: Conduct a risk assessment for the survey.

Step 1: Collect all the relevant information to plan the survey

74 It is essential that the surveyor collects all the necessary relevant information to ensure that the survey is completed efficiently, effectively and safely, and that it meets the client's requirements. The information should be gathered as early as possible to enable thorough planning. The ideal situation would be to arrange a preliminary site meeting and have a walk-through inspection. This is essential for large and complex premises. However, pre-survey site visits may not always be possible (eg small surveys where the cost of a second visit outweighs advantages or where there are multiple premises (eg chain stores) and it is not practical to visit them all). In such situations the information will need to be gathered through other means (eg by correspondence such as phone/e-mail/post, or by a preliminary meeting and walk-through immediately before the survey).

75 The information required is listed in Box 7. It is often easier to obtain this information through direct discussion with the client. The meeting is also an opportunity to explain further the nature of the survey and material assessments and agree the nature and format of the results and report.

76 Accurate plans of the building(s) and the floor layout should be obtained at this stage where possible. Building plans should be used for complex premises. The plans should contain the main features of each room, corridors, stairs etc. The plans should be marked with unique floor and room numbers to help identify individual locations. The plans should be checked for accuracy and completeness. If plans are not available, an accurate drawing of the premises will need to be made by the surveyor before the survey starts. In some premises (eg small/uncomplicated), a simple drawing showing the salient features may be sufficient. In other situations, more detailed drawings will have to be made. These plans will be used to refer to and record the position of any suspect material and the location of any samples taken for identification. The plans should also be used to locate and record any sensitive or restricted areas and hazards.

Box 7: Information to be collected by the surveyor

- Description and use of property (ie industrial, office, retail, domestic, education, health care etc).
- Number of buildings: age, type and construction details.
- Number of rooms.
- Any unusual features, underground sections.
- Details about whether the buildings have been extended, adapted or refurbished, and if they have, when the work was done.
- Any plant or equipment installed.
- Whether a listed building, conservation area etc.
- Extent or scope of survey required (possibly mark details on a site plan or architects' drawings).
- Whether the surrounding ground and associated buildings or structures are to be included in the scope of the survey.
- Current plans or drawings of the site.
- Previous plans, including architects' original drawings and specifications and subsequent plans for major changes and refurbishment.
- Whether the premises are vacant or occupied.
- Any restrictions on access.
- Special requirements or instructions.
- Responsibility and arrangements for access.
- Whether survey damage is to be made good (refurbishment/demolition surveys).
- Site-specific hazards (mechanical, electrical, chemical etc).
- Responsibility for isolation of services, power, gas, chemicals etc.
- Working machinery or plant (including lifts) to be made safe (these are covered in greater detail in Step 4).
- If photos are to be taken.
- How many bulk samples will be necessary.
- The location of all services, heating and ventilation ducts, plant rooms, riser shafts and lift shafts.
- Details of any previous asbestos surveys (Type 1/2/3 Surveys), current asbestos registers and all records of asbestos removal or repairs.
- Information on possible repairs to ACMs, eg pipe/thermal insulation.
- History of the site: any buildings previously demolished; presence of underground ducts or shafts etc.

Preliminary site meeting and walk-through inspection

77 A walk-through inspection will be extremely valuable for planning the survey and identifying potential issues and problems. The inspection will enable the surveyor to become familiar with the layout of the premises, including the location of equipment and furniture etc, which may impede access or sampling. In addition, it will allow the surveyor to gain an appreciation of the size of the project and to estimate the extent of sampling required. The inspection also allows any specific hazards to be recognised and discussed to minimise the risks. It will also enable other potential issues to be identified or raised (and resolved). Possible issues include entry or access restrictions (eg to ceiling voids, high areas and crawl spaces), sampling matters (eg sampling only when the area is unoccupied, materials or decorations which cannot be disturbed, labelling of sample locations, future placement of asbestos warning labels (see Figure 3), measures used to reduce dust release and clean-up etc) and the potential need for a licensed asbestos contractor (eg to gain access through AIB ceiling tiles). Where access is required to high areas, arrangements should be made for the use of scaffolding, a tower crane or MEWP. The walk-through inspection should also be used to check the accuracy of the building plans.

78 If a pre-survey site meeting and walk-through are not possible, the information listed in Box 7 should be collected through discussion and correspondence with the client. Any information not collected will need to be obtained at the site prior to the survey starting. In addition, plans and drawings will need to be checked at this time.

Step 2: Consider the information (desk-top study)

79 The surveyor needs to collate and consider all the collected information (eg on the premises, building structures, processes, plant and machinery types) so the survey can be properly planned. This is a 'desk-top' exercise to review the information, plan the survey strategy and consider if there are any gaps in the information. The surveyor should also consider the resources and equipment etc, which will be necessary to complete the work. The surveyor should consider the information on the following:

- competency to undertake the work;
- available resources;
- intended programme of works;
- expected equipment to be used for access:
  - into the structure;
  - to high levels;
  - into contaminated areas or confined spaces;
  - through known ACMs;
- the need for additional trades (joiner, electrician, builder) to gain access during the survey or to reinstate areas on completion;
- bulk sampling strategy and expected number of samples to be taken with reference to the site plan.

80 Many premises will be relatively simple and straightforward, eg one or two buildings and no additional land or ground, no machinery, lifts and outbuildings and no previous refurbishment or demolition. The 'desk-top' review is the time to focus on the nature of the premises and the type of survey. Refurbishment and demolition surveys in particular will need much consideration.

Step 3: Prepare a survey plan (including how data will be recorded)

81 After all the relevant information has been collected and the preliminary site inspection and desk-top study have been completed, a written plan for the main survey can be produced. The plan essentially sets out the content of the survey and can form the basis of the contract with the client. The plan will normally specify the following:

Figure 3 Asbestos label

Scope
- The scope of the buildings' survey.
- Any external areas to be included.
- Any areas to be excluded.
- The type of survey (management or refurbishment and demolition).
- Any possible or known ACMs, **not** to be included in the survey.

Survey procedure
- The survey procedure (eg how it will be conducted) and sampling strategy including:
  – agreed numbers of samples and sampling methods;
  – agreed numbers of photographs;
  – procedures for making good;
  – agreed survey times of work;
  – agreed signage;
  – key access;
  – agreed start and completion dates;
- The material assessment method and the parameters to be assessed (eg product type, location, extent, condition and accessibility of ACMs).
- The information to be recorded and the method and format to be used.
- The quality assurance checks and procedures to be undertaken.
- Any known area where access will not be possible.

Personnel and safety issues
- Names of surveyors (for security purposes).
- Safety precautions from the surveyor's risk assessment, including steps to minimise asbestos disturbance and prevent asbestos spread.
- Site safety procedures for emergencies including decontamination etc.

Report
- Report format with headings (see paragraph 132).
- What data will be reported.
- How the data will be presented (each room/area should be individually recorded).
- The way the survey data will be stored, accessed and updated (eg a paper copy in the site manager's office or a computer database accessible on a network or the internet).
- The way photographic or video records and marked-up plans will be stored and reported.
- How to record asbestos look-alike materials (if not sampled).
- Other information required by the client that may have been agreed, eg detailing fixings.

82 The survey report should contain a summary of the results in a format that can be used as the basis for an updatable register of ACMs (ie the asbestos register) and a diagram (ie building drawings) indicating the locations of ACMs (see paragraphs 131–144). This register will need to be readily accessible to all involved in initiating maintenance or other work on the fabric of the building. It should be available in hard copy format and, where appropriate, stored electronically.

Step 4: Conduct a risk assessment for the survey

83 Surveying will present health and safety issues to the surveyors and others. Therefore before a site survey, it is important that an assessment of the risks to the health and safety of surveyors, sampling personnel and building occupants is carried out. The client should provide information relating to any hazards specific to the site at the Step 1 stage. The types of non-asbestos hazards which may be associated with surveys include:

- working at heights, in ceiling voids or on a fragile roof;
- working on operable machinery or plant;
- working in confined spaces;
- chemical hazards;
- electrical hazards;
- biological hazards;
- noise hazards; and
- lone working.

84 There may also be other specific hazards in certain types of premises, eg hospitals and nuclear plant have radiological hazards.

85 The risk assessment should be prepared by a competent person (normally the surveyor) and it should be written down. It should establish all the hazards at the particular premises and go on to identify the correct precautions and procedures in a **plan of work** for the survey. In many cases, surveyors will only see the site for the first time at the survey, so they will have little chance to evaluate the site-specific hazards that are involved and will rely on the risk assessment made based on information collected at Stage 1. The risk assessment should also specifically address the asbestos issues, including:

- the need to prevent disturbance of ACMs as far as possible;
- the need to prevent the spread of ACMs;
- identification of safe work procedures (eg controls to be used while taking samples, arrangements for entering contaminated areas);
- PPE to be used;
- decontamination and disposal arrangements.

86 Information on safe systems of work for asbestos sampling is set out in paragraphs 110–111.

87 Refurbishment/demolition surveys are more likely to present some serious health and safety hazards due to the intrusive and destructive nature of the work, eg hidden electrical cables or pipes or unstable buildings. The hazards will need to be properly addressed with procedures in place to deal with emergencies.

Safe work procedures
88 Ideally a survey should be conducted with a team(s) of two people. This has a number of advantages, for example in assisting with carrying equipment such as step ladders, labelling of sample bags and documentation. In cases of remote or dangerous locations (eg derelict buildings or items identified in paragraphs 83–84), a team of two should be specified as a safety requirement. Team working also allows field training of new surveyors to be carried out in a supervised practical environment and gives a better chance of finding ACMs. Further information on safe working procedures can be found in paragraphs 110–111.

Personal protective equipment
89 Disposable coveralls, overshoes and gloves should be worn when there is a likelihood of asbestos contaminating the surveyor's clothing and during bulk sampling. The coveralls should be the type normally used for asbestos work (ie Type 5 coverall) and should have a hood and elasticated cuffs and ankles. They can usually be worn over normal clothing, but should be carefully removed after use by turning inside out, and be disposed of as asbestos waste. Coveralls should not be reused after they have been taken off. The coveralls are usually rolled inside out to minimise spread so that the outside makes contact with the inside: if reused they will contaminate normal clothing. Take care to prevent the spread of asbestos. For some dirty or contaminated sites, Wellington boots will be required, and these should be wiped or washed clean if they become contaminated. They should also be cleaned after sampling is completed. For some sites, more stringent decontamination procedures may be needed (see paragraph 91). **There should be an appropriate statement in the generic plan of work as to what type of situation will trigger this, so that even at sites where a preliminary meeting etc was not feasible, appropriate precautions can be taken by the surveyor.**

90 Appropriate RPE should be worn during sampling or when surveying areas where the asbestos is likely to be disturbed during the inspection (eg crawl tunnels and above false ceilings). The survey and sampling personnel must have been properly trained in the selection, use and maintenance of RPE and follow the guidance given in *Asbestos: The analysts' guide for sampling, analysis and clearance procedures*.[15] In many cases a disposable FFP3 respirator or a half mask fitted with a P3 filter will provide adequate protection. Face-fit tests should be carried out to confirm that the mask fits the wearer.

Decontamination and disposal arrangements
91 If the surveyor has to enter areas where there is significant contamination (eg thermal insulation in crawl tunnels, spray insulation in ceiling voids), there is a greater potential for contamination of clothing and footwear. The risk assessment must take these conditions into account, as additional safety precautions and decontamination procedures will be needed. It may involve a higher standard of personal protection (eg powered full-facepiece respirator fitted with a P3 filter) and more comprehensive decontamination procedures (eg use of a decontamination unit). Where entry into these locations is necessary, surveyors must be adequately trained in the use of high-performance RPE and in decontamination procedures (decontamination procedures are covered in *Asbestos: The analysts' guide for sampling, analysis and clearance procedures*). Surveyors should not wear their own clothes under coveralls in these circumstances. In addition, there should be appropriate discussion between the surveyor and client to ensure the relevant decontamination procedures are employed. If significant contamination is unexpectedly encountered in the course of the survey, then 'emergency' procedures should be implemented, eg leave the area and discuss with the client.

5 Carrying out the survey (surveying)

Introduction

92 Surveys should be carried out methodically, systematically and diligently to make sure ACMs are not missed and all areas of the premises are inspected. Building plans should be used to prepare the survey strategy and for checking progress through the premises. Plans should be inspected to make sure building features and services (eg voids, cavities, risers, ducting, undercrofts etc) are included. There are various options for a systematic survey inspection. One example is shown in Box 8.

93 Each area should be surveyed with due care to avoid missing any ACMs. Surveyors should be inquisitive and use initiative. Materials should be tapped and prodded. Everything should be checked and inspected. Do not presume every item is the same just because it looks similar. This is particularly relevant when assuming items are non-asbestos. Sample and take photographs as you go along. Look out for unusual, potential sources such as overspray or packers. Allow enough time for the survey. It is good practice to survey 'in pairs' ie two people working together, with both inspecting one area at the same time. Recheck areas which are complex or have many items (eg plant rooms). ACMs will be missed where surveyors are tired, rushed or make assumptions. Do a final walk through, checking notes against plans. Large premises will require more detailed survey procedures, particularly if several surveyors are involved, eg it may be appropriate to carry out a separate survey on the building services, machinery and any large floor and ceiling voids; and recaps and checks should be carried out frequently.

Box 8: Example of a systematic survey inspection

External areas:
- Work downwards from high to low.
- Work from the periphery inwards.

Internal areas:
- Work upwards from basement to roof.
- Inspect each area individually.
- Work around each area clockwise from the door of entry.
- Inspect each component inside each compartment in the following order: ceiling, walls, floors, fixtures and fittings, equipment and services.
- Look at each item individually.

General:
- Check and inspect everything.
- Sample and take photographs as you go along.
- Recheck areas which are complex or have many items.
- Do a final walk-through, checking notes against plans.

Types, location and appearance of asbestos-containing building products: Appendices

94 Appendices 2–3 provide detailed information on ACMs in buildings. Appendix 2 summarises the main types and uses of ACMs in the fabric of a building and in fixed installations such as heating, water and electrical systems. It lists the main product types, their location and use, asbestos content, date last used and common trade or product names. The product types are listed approximately in order of their ability to release fibres assuming no surface treatments have been applied. The locations of many of these products in buildings are shown diagrammatically in Appendix 3 together with an extensive picture gallery of many asbestos products as an aid to identifying ACMs.

Older industrial machinery and plant

95 Older equipment is likely to contain asbestos due to its age or higher performance requirements. The equipment is also likely to need servicing and maintenance. The surveyor should inspect the accessible parts of machinery and plant which provide heat and electrical insulation, high-performance seals and frictional performance (eg driving belts, clutches, brakes and bearings). The surveyor should not sample or work on any machinery unless qualified to do so. Engineers or maintenance personnel may be able to help in these situations. If sampling is not carried out, the equipment should be presumed to contain asbestos unless there is evidence that it contains non-asbestos materials.

Older consumer electrical products

96 Older consumer-type industrial electrical products may also contain some ACMs (eg hairdryers, irons, washing machines, dishwashers, tumble dryers). However as the asbestos in this type of equipment is not readily accessible and presents only a very low risk, it is not practical to inspect or sample for it. However, products which are used for or require significant heat insulation should be inspected during the survey. These will include simmering mats, iron stands, fire curtains and blankets, catalytic gas heaters, all types of warm air, storage or radiant heaters, and cooker door seals.

What to assess and record

97 The management survey requires that the condition of the ACMs and their ability to release fibres are assessed (ie a material assessment). Therefore, for a management survey, the information listed in Box 9 should be obtained and recorded for each ACM or presumed or suspect ACM. Refurbishment and demolition surveys normally require less information as details of ACM condition are not required (see Box 10 but note time reference).

> **Box 9: Information required for a management survey**
>
> - Asbestos product type(s).
> - Location of the material(s).
> - Extent (or quantity) of the material(s).
> - Asbestos type(s).
> - Accessibility and/or vulnerability of the material(s).
> - Amount of damage or deterioration.
> - Surface treatment (if any).

> **Box 10: Information* required for a refurbishment or demolition survey**
>
> - Asbestos product type(s).
> - Location of the material(s).
> - Extent (or quantity) of the material(s).
> - Asbestos type(s).
>
> * Where the refurbishment or demolition work will not take place for a significant period after the survey (eg three months), then the information required for a management survey should be obtained (see paragraph 52).

Presumed ACMs

98 If a sample is not taken, there must also be a presumption made whether the material is asbestos or non-asbestos. Surveyors may visually assess the edges and damaged areas of suspect materials and record the following:

- whether visible fibres are present on close inspection (see *Asbestos: The analysts' guide for sampling, analysis and clearance procedures*);
- the colour of the fibres, if visible; and
- whether fibres are visually consistent with asbestos (eg form bundles with splayed ends).

99 Some materials, like textured plasters, paints and vinyl floor tiles, may contain very fine dispersed chrysotile asbestos which may not be seen by eye or with a magnifying glass, and these materials (if old) will have to be presumed to contain asbestos unless they are sampled and carefully analysed by a competent laboratory. As imported materials may have contained chrysotile asbestos until 1999 and laboratories often miss the fine asbestos, some additional checks may be necessary with these types of materials. Other useful characteristics (eg surface texture, sound when knocked, warmth to touch, surface hardness/deformation with a probe) may also be used by experienced surveyors to help compare the material with other materials they have previously encountered and had samples identified. Unless the surveyor is convinced that there is adequate evidence to conclude that the material is asbestos-free (eg plaster, plasterboard, wood etc), a presumption or strong presumption should be made that it is an ACM.

Asbestos type

100 The material assessment requires information on the type of asbestos. In the absence of analytical data and where similar products have not been identified in the survey, the most likely asbestos type must be allocated based on the product types and age in Appendix 2. It may also be possible to obtain information on the type of asbestos from close inspection of the material, eg if fibres are visible in the product, these can give some additional clues to the type of asbestos (see *Asbestos: The analysts' guide for sampling, analysis and clearance procedures*). In general, however, unless there is evidence to show otherwise, the asbestos type should be assumed to be crocidolite asbestos.

Bulk sampling strategy

101 Sampling will normally be carried out at the time of the survey. This is usually the most convenient and efficient arrangement. However, for very large premises or where access has not been possible, sampling may be carried out as a separate exercise, eg when the area is available. Each area and room in the premises should have a thorough visual examination to identify the materials and locations to be selected for sampling. The visual inspection should be conducted systematically (see paragraph 93 and Box 8). Materials should be inspected for apparent differences and variation in appearance. Samples of about 3–5 cm^2 surface area and through the entire depth of the ACM (including any backing paper) should normally be taken with the aim of collecting one or more samples which are representative of the whole material. Sampling should not be carried out where there is an electrical hazard or if it will damage the critical integrity of a roof, gutter, pipe etc. An equipment checklist for sampling is given in Appendix 5.

102 The sampling strategy will be based on several factors, including the size and numbers of premises/rooms and the extent, types and variation in materials present. The visual inspection and checking (eg tapping and prodding) of each material will allow the sample numbers and locations to be specified. In general, for homogeneous manufactured products containing asbestos, it can be assumed that the asbestos is uniformly distributed throughout the material, and one or two samples will suffice, eg boards, sheets, cement products, textiles, ropes, friction products, plastics and vinyls, mastics, sealant, bitumen roofing felt and gaskets. Insulation materials are generally less homogeneous as they were applied on site and their composition depended on the availability of supply. Subsequent repairs and patching may add to this variability and increase the number of samples required. Repaired and replaced materials should always be sampled in addition to the original items.

103 In addition, substantial contamination and debris may have been produced at the time of installation, eg overspray, other insulation debris, AIB off-cuts. A favourite practice was to drop off-cuts into voids and sweep debris into lift shafts and other risers. Asbestos debris and other suspect visible contamination should be sampled.

104 For homogeneous material, often a single sample may be all that is required to confirm the suspicion that it is asbestos and to make a presumption that it applies to other material of the same type. However, for non-homogeneous materials and for some presumed non-asbestos materials, additional sampling may often be needed, to reduce the possibility of false negatives which may lead to incorrect conclusions. The following sample numbers are suggested for each room or defined area, but may be adapted depending on the site and the circumstances prevailing.

Spray coatings, encapsulated sprays and bulk materials

105 These are usually, but not always, homogeneous (under any encapsulate). Different mixtures may have been used and material may have been removed, repaired or patched at various times. Where the material appears uniform and consistent, two samples should usually be enough, if taken at either end of the sprayed surface. If the installation is particularly large (eg >100 m^2), one sample should be taken approximately every 25–30 m^2. Samples should be taken from all patches of repairs or alterations.

Pipe/thermal insulation

106 Pipe insulation is often highly variable in composition, especially where there is a change in colour, size and texture or there is evidence of repairs or modifications, eg asbestos may have been stripped from long runs of pipes but have been left around pipe elbows, taps and valves. In general, one sample should be taken per 3 m run of pipe with particular attention paid to different layers and functional items (valves etc). For long runs of pipes (eg >20 m), one sample per 6 m will usually be enough. It can be difficult to demonstrate that individual pipes are asbestos free so all pipes should be sampled even when they appear similar (see Figure 4). Samples should be taken from all patches of repairs or alterations.

Figure 4 Pipe uncertainty: Top pipe is non-asbestos. All pipes should be sampled

Insulating board

107 Insulating board is usually homogeneous but repairs and replacement boards and tiles may have been fitted. Boards and tiles may also have been painted. One sample per room or every 25 m^2 is usually adequate. Ceilings and walls should be thoroughly inspected to check for variation and differences. Many premises have had individual or groups of AIB tiles replaced as part of improvement programmes. If there is evidently more than one type of tile (eg based on colour, pattern, design, size etc) then representative samples of each should be taken. Larger installations completed at the same time may require only a few tile samples to be taken. Some replacement tiles may look the same. Inspection of the hidden side of the board or tile may, where access permits, reveal the trade name of the materials and/or differences in colour which indicate variations in the material. Insulating board or tiles may occasionally have been manufactured with asbestos paper on one or both sides.

Asbestos cement materials

108 These are homogeneous materials which are commonly encountered as corrugated and flat sheets or as various moulded products. Asbestos cement (AC) was also widely used in low-cost housing as wall and ceiling panels and in schools in fume cabinets and kick boards. It was also mistakenly used as fireproofing and therefore is found in places where AIB is expected, as well as on office partition walls. It does not always look like AC in these situations. In older buildings, most pre-formed exterior cement sheets can be strongly presumed to be asbestos and only limited sampling is needed to confirm the presumption. The risk from falls through asbestos cement roofs usually means that sampling is restricted. If sampling is required, one sample of each type of sheet or product (eg gutters, downpipes etc) should be taken. Repeated sampling is not usually necessary unless areas of replaced sheets are found. Asbestos cement sheets are visually very similar to their non-asbestos (fibre cement) replacement. Fibre-cement replacement sheets are identifiable by a code 'NT' placed near the edge of the sheet, where they overlap. Some 'newer' asbestos sheets have the code 'AT' in a similar position.

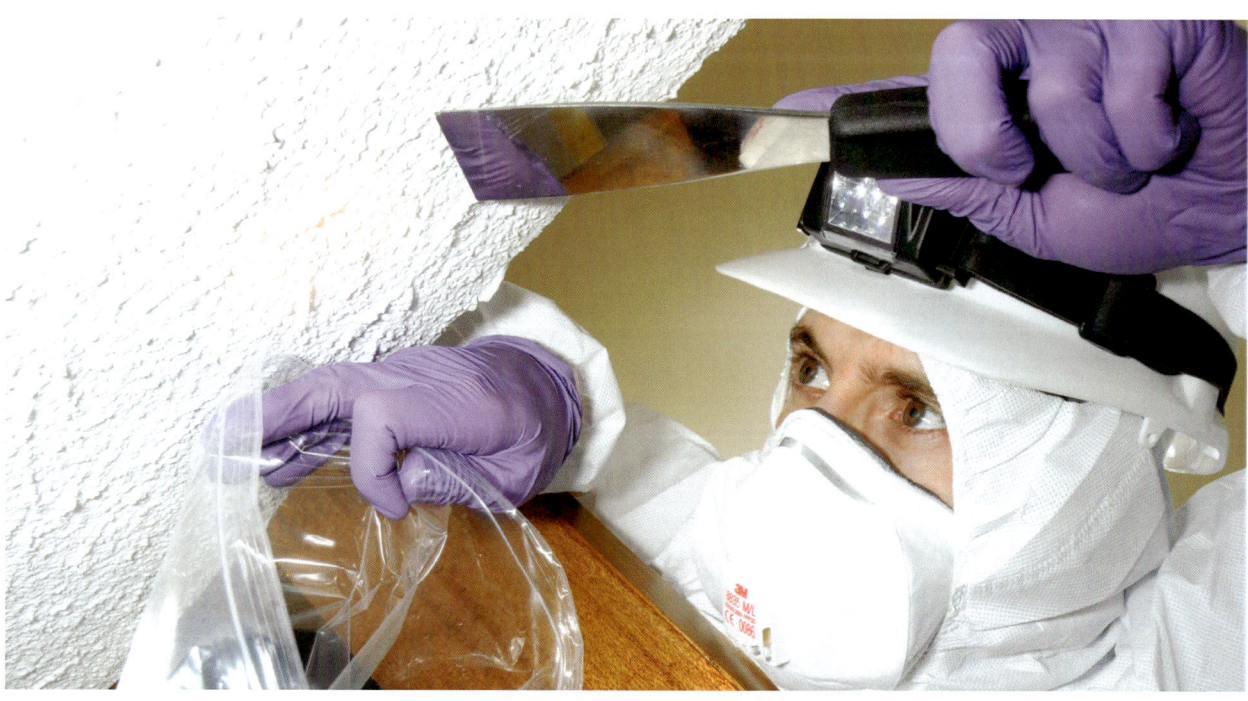

Other materials including debris and contamination

109 Where there are distinct types of materials, then one or two samples from each separate source will usually be adequate. Two samples are recommended if there are more than a few square metres of material.

Bulk sampling procedures

Safe systems of work

110 All work to be carried out must have an adequate risk assessment of the survey site (see paragraphs 83–87) and the work must be carried out according to the procedures defined in the risk assessment. The work should minimise the disruption to the client's operations and must protect the health and safety of everyone who may be at risk. Sampling personnel must wear adequate PPE (see paragraphs 89–91). Airborne emissions should normally be controlled by pre-wetting the material to be sampled, with water and/or a suitable wetting agent. This may involve spraying the surface (eg boards and sheets) or injecting (eg lagging and sprays). Shadow vacuuming (holding the suction inlet close to the area where dust is being produced) with a Class H (BS EN 60335)[16] vacuum cleaner should be used if wetting is likely to be incomplete (eg AC sheets, AIB boards, ropes and gaskets) or if it is not safe to do so (eg it may drip into electrical installations). Special sampling precautions are used for pipe insulation (see paragraphs 114–115).

111 The areas to be sampled inside buildings should as far as possible be unoccupied. Sampling should not be undertaken in normally occupied areas, but if in constant use, periods of minimal occupation should be chosen. The nature of the area, the likely release of dust and the proximity and nature of future work will dictate the precautions required to prevent the spread of asbestos. Entry of other people to any sampling area should be restricted or suitable warnings posted (eg a notice with wording such as 'Asbestos sampling in progress: Keep out'). Care should be taken to minimise disturbance to ACMs and any dust or debris that might be present. Surfaces onto which asbestos debris may fall should be protected with a sheet of impervious material such as polythene which can be easily cleaned by wet-wiping or using a suitable Class H vacuum cleaner. All samples must be individually sealed in their own container or a sealable polythene bag which is then sealed in a second container or polythene bag. The sample area should be left clean with no evidence of debris from the sampling operation and any sampling points sealed to prevent the release of fibres. Various methods are used to reseal the sampling point (eg tapes and fillers).

Sample and site labelling

112 Whenever a sample is collected, it should be labelled with a unique identifier that is also recorded in the survey documentation, records and site plans so that the sample origin can be traced at a later date. The sampling position at the site may also be labelled with the same identifier. Visual records such as marked-up plans and/or photographic records showing the location and extent of the sample are also effective ways of recording the sampling position and the location of the ACMs.

Bulk sampling

Spray coatings and bulk materials

113 If the coating is encapsulated, it can be pre-injected with liquid around the sampling area then carefully cut with a sharp knife or scalpel to lift a small flap to retrieve a sample. If the spray coating is not covered, both wetting (spraying surface and injection) and shadow vacuuming may be necessary to reduce airborne emissions. As spray coatings are generally homogeneous, a surface sample should suffice.

Pipe insulation

114 The area to be sampled should be fully wetted first: injection techniques are recommended. Samples are taken with a core sampler which should penetrate to the full depth of the pipe insulation. Proprietary types are available, but laboratory cork borers are also suitable. It should include a plunger to remove the sample from the borer. The sample point hole should be made safe after sampling (eg covered with tape or filled with a suitable inert filler), if the pipe is to remain in place and the surface was originally intact. This helps to keep the insulation in good condition and to prevent dispersal of asbestos. The borer should have a wet wipe pushed down to form a plug inside the borer and another wrapped around the outside. The borer is then used to take a full-depth sample of the insulation. The inner wet wipe is used to seal the surface of the insulation where the borer enters and disturbs the insulation. The outer wet wipe is used to clean the outside of the borer as it is withdrawn, and the contaminated wet wipe can be placed in the sample bag. The sample is removed by using the plunger to push the sample out into the polythene bag, complete with the wet wipe. Further cleaning will be required to completely clean the sampling equipment between samples.

115 An alternative approach is to use core sampling tubes in which the sample is retained. Again the core tube can be withdrawn through a wet wipe and then capped at both ends and placed in a bag until it reaches the laboratory. Chicken wire was often included within pipe insulation. This may hamper sampling, and a thin core sample may need to be taken. Where there is pipe insulation which is obviously new and non-asbestos, the possibility of debris from an earlier asbestos strip beneath the new insulation should be investigated.

Insulating board
116 Materials such as ceiling tiles or wall panels should be inspected for areas of existing damage, where a sample can be collected more easily. Otherwise, a small sample should be taken from a discrete location at the corner or edge of the panel, with a sharp knife or chisel blade to lever off a sample. Make sure that any paper, on one or both sides, is included.

Asbestos cement
117 Asbestos cement can usually be identified by visual inspection. Where sampling is necessary (eg to distinguish between AC and AIB), look for a damaged portion where it will be easier to remove a small sample (AC is usually very hard). The sample size should be about 5 cm^2 as it will be necessary to search for traces of amphibole asbestos such as crocidolite and amosite. The sample should be obtained using pliers or a screwdriver blade to remove a small section from an edge or corner. **Samples should not be collected from roofs without special safety precautions to prevent falls through the fragile sheets.** If the analysis is still inconclusive (eg chrysotile and amosite are detected), then the definitive water absorption test should be conducted (the material will be classed as AC if it absorbs <30% water) (see *Work with materials containing asbestos*).

Gaskets, rope, seals, paper, felts and textiles
118 Samples can be taken using a sharp knife to cut a representative portion from the material.

Floor and wall coverings
119 Samples should be cut out with a sharp knife, usually taking one sample from tiles of each type or colour present. The area should be cleaned after sampling but the fibre release is likely to be very low, unless the asbestos is present as a lining or backing material.

Textured coatings
120 Samples should be obtained by carefully scraping the coating with a screwdriver or narrow scraper, directing the material into the sample container held below the sampling point.

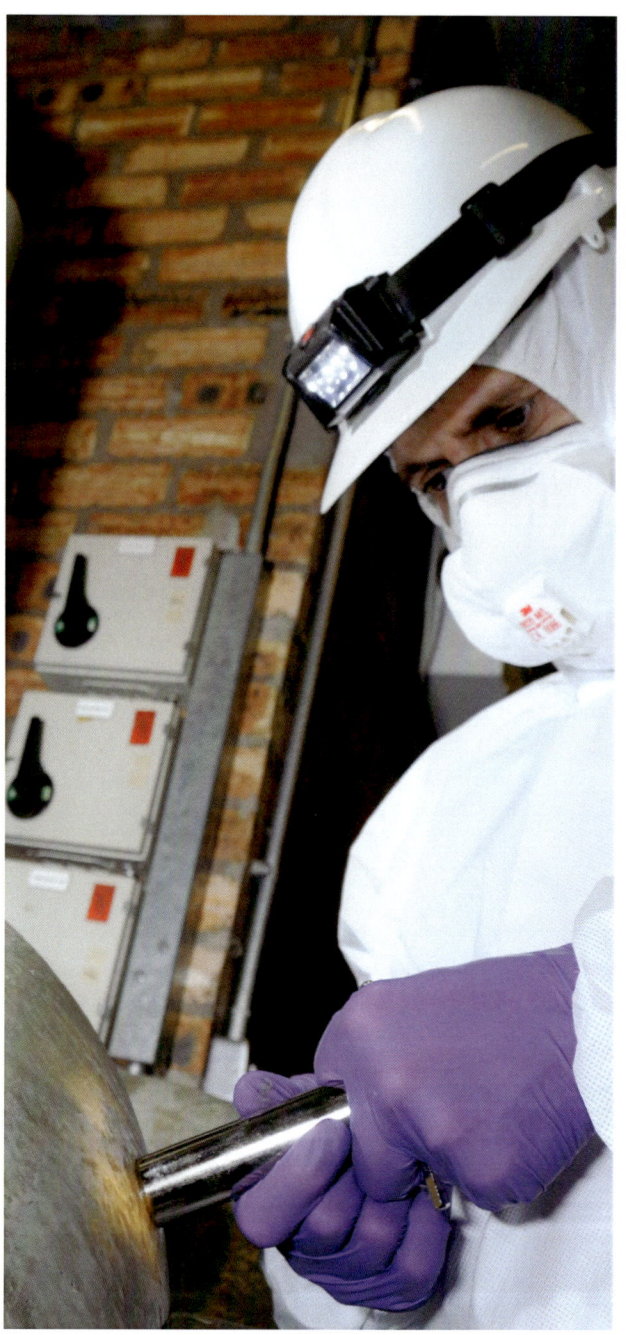

Air sampling
121 Personal air sampling can be carried out to measure the exposures of survey and sampling personnel. Occasionally there may be a request for 'background' air sampling if the ACMs are a matter of sensitivity to the occupants. Such requests need careful appraisal, as the area may already be contaminated, even before the bulk sampling is carried out. Air sampling may also be required where there has been intrusive sampling (eg in refurbishment or demolition surveys) and areas or buildings are to be reoccupied for a period before the work is carried out (see paragraphs 53–54). The procedures for reassurance air sampling as described in *Asbestos: The analysts' guide for sampling, analysis and clearance procedures* should be used.

Sample analysis and reporting

122 Analysis of the samples collected should be carried out and reported in accordance with the method given in *Asbestos: The analysts' guide for sampling, analysis and clearance procedures* or an equivalent method. The laboratory report should for each sample give a clear statement of whether asbestos was found and the types of asbestos identified.

123 Laboratory results should be appended. Materials which have been sampled and found not to contain asbestos after analysis also need recording as **asbestos not detected**, as the asbestos content of these materials may be questioned in future and it will save a great deal of time and cost if this has been clearly recorded in the first instance. The survey report or abstracts from it should be presented in a form that can be used as the basis of a register or log of ACMs (see paragraph 139).

Material assessment

124 As outlined in paragraph 44, the management survey should include an assessment of the condition of the various ACMs and their ability to release fibres. This assessment allows the dutyholder to assess the potential for fibre release for each ACM and then go on to prioritise the need for action as part of the plan for managing asbestos. **The material assessment should be carried out as part of the management survey.** A standardised assessment tool suitable for a management survey is given in paragraphs 125–127. It is based on a simple additive algorithm. The tool can be used to numerically assess the potential for fibre release. The tool is not designed to calculate absolute differences in potency or fibre release/hazard potential between ACMs. It does however enable ACMs to be ranked in a simple numerical order. No condition assessment is normally necessary for refurbishment and demolition surveys but, where the period between survey and the event is significant, eg more than three months, then a material assessment should be conducted and interim management arrangements put in place.

Material assessment algorithm

125 In the material assessment process, the main factors influencing fibre release are given a score which can then be added together to obtain a material assessment rating. The four main parameters which determine the amount of fibre released from an ACM when subject to disturbance are:

- product type;
- extent of damage or deterioration;
- surface treatment; and
- asbestos type.

126 Each parameter is scored between 1 and 3. A score of 1 is equivalent to a low potential for fibre release, 2 = medium and 3 = high. Two parameters can also be given a nil score (equivalent to a very low potential for fibre release). The value assigned to each of the four parameters is added together to give a total score of between 2 and 12. Presumed or strongly presumed ACMs are scored as crocidolite (ie score = 3) unless there is strong evidence to show otherwise. Examples of scoring for each parameter are given in Appendix 4.

127 Materials with assessment scores of 10 or more are rated as having a high potential to release fibres, if disturbed. Scores of between 7 and 9 are regarded as having a medium potential, and between 5 and 6 a low potential. Scores of 4 or less have a very low potential to release fibres. Non-asbestos materials are not scored.

Risk assessment and management plans

128 The material assessment identifies the 'high-hazard' materials, ie those materials which will most readily release airborne fibres if disturbed. It does not automatically follow that those materials assigned the highest score in the material assessment will be the priority for remedial action. Priority must be determined by carrying out a risk assessment (ie a priority assessment) which will take into account factors such as:

- the location of the material;
- the extent of the material;
- the use to which the location is put;
- the occupancy of the area;
- the activities carried on in the area; and
- the likelihood/frequency with which maintenance activities are likely to take place.

129 The priority assessment can only be carried out with the detailed knowledge of all these factors. The surveyor can help in this process, by obtaining information which will contribute to the priority assessment, particularly in small or simple premises where information on occupancy and use is straightforward. However, such help must be undertaken with caution. It is the dutyholder, under CAR 2012, who is required to make the risk assessment using their detailed knowledge of the activities carried out in the premises.

130 The combined material and priority assessment results should be used to establish the priority for those ACMs needing remedial action and the type of action that will be taken. There are various remedial options available: in many cases the ACMs can be protected or enclosed, sealed or encapsulated, or repaired. These options should be considered first. Where such actions are not practical, ACMs should be removed.

6 Survey report

131 The survey report is a record of the information collected at a particular time on the presence and condition of ACMs. Extreme care and attention should be paid to producing the report, particularly in transposing data, as the document will be the formal record of the survey. It will contain the information and data that will be used to prepare the risk assessment and management plan and to make decisions and judgements on the need for actions. Errors in the report could lead to incorrect conclusions and inappropriate decisions.

132 The report should be completed in a written format, supplied either as a hard copy or as an electronic document, or both. It should be comprehensible to and usable by the client. In particular, the information in the survey report should be easy for the client to extract and to use to prepare an asbestos register, eg by presenting the results in a manner or format that can be directly lifted or employed to form the asbestos register. The report should contain the results of sample analyses. The survey report should contain the following sections:

- executive summary;
- introduction covering the scope of work;
- general site and survey information;
- survey results (including material assessment results);
- conclusions and actions;
- bulk analysis results.

133 The design, layout, content and size of the report are very important. Large reports can be unwieldy and even intimidating. Clients are most interested in the summary, results, conclusions and actions. In hard-copy documents, it can be useful to separate the report into different parts, with the bulk analysis results and the individual survey results, particularly if displayed with accompanying photographs, contained in separate detachable appendices.

Executive summary

134 The executive summary should briefly describe the scope, type and extent of the survey and it should summarise the most important information, including:

- the locations with identified (or presumed) ACMs;
- areas not accessed (which should be specific to the survey and not generic);
- ACMs with high material assessment scores;
- clear notes on any actions (and priorities).

Introduction

135 The introduction should explain the scope of the work and the purpose, aims and objectives of the survey. It should also contain a description of the nature and age of the building(s) (or other structures) plus construction type.

General site information

136 General site and survey information should include:

- the name and address of the organisation;
- the names of the surveyors;
- the name and address of the person who commissioned the survey;
- the name and address of the premises surveyed;
- the date of the report;
- the date of the survey;
- a description of the areas included in the survey;
- a description of any areas excluded in the survey;
- the survey method used (this publication and/or other documented procedures);
- the type of survey undertaken (management or refurbishment/demolition) and, if more than one type is used, where they apply within the premises;
- any variations or deviations from the method; and
- agreed exclusions and inaccessible areas (with reasons) which should be specific to the survey and not generic.

Survey results

137 The survey results should be summarised in table format and as a set of marked-up plans (diagrams) showing the location of ACMs and presumed ACMs. The summary table should contain the following information:

- location of the ACMs (eg building identifier, floor

number or level, room identifier and position);
- extent of the ACMs (area, length, thickness and volume, as appropriate);
- product type (see Appendix 2);
- level of identification of the ACM (presumed, strongly presumed or identified); and
- asbestos type in the ACM (eg chrysotile, amosite, crocidolite).

138 For a management survey (and refurbishment and demolition surveys where the work is not imminent), the following additional information should be provided:

- accessibility of the ACM;
- amount of damage or deterioration;
- surface treatment (if any);
- the material assessment score or category (high, medium, low or very low);
- any actions required from the material assessment.

139 Table 1 shows the presentation of survey results. This format can be easily incorporated directly into the asbestos register. Figure 5 shows an example of a marked-up building plan.

140 The information in the results table should be presented on an individual room basis. Any rooms or areas not accessed and presumed to contain asbestos should be included in the results table. If a priority assessment has been included, the priority scores should be listed and any actions required highlighted. (Note: the priority assessment should only be carried out in consultation with the client or dutyholder, who must provide accurate information on all the activities carried out on the premises.)

141 Where suspect material is proved not to be asbestos, by sampling or other means, this should be recorded in a separate table. This will help in any future debate over the nature of these materials.

Address: **Date:**

Location	Product type	Extent	Accessibility	Condition	Surface treatment	Asbestos type	Sample no	Sampled/ presumed/ strongly presumed	Material assessment score and action	Priority score Action
Store room 2, BC408, ceiling	AIB	Whole ceiling 120 m^2	Medium	Good	Painted one face only	Amosite	1	Sampled 4 samples	5	12
Store room 2, BC408, fire door	Asbestos board on door carcass (AIB)	21 m^2	Medium	Good	Encapsulated by wood in door	Amosite	2	Sampled 1 sample	5	12
Meeting room 2, BC412, ceiling	Asbestos ceiling tiles (AIB)	5 m^2	Medium	Good	Painted one face only	Amosite	3	Sampled 1 sample	5	13
Canteen, BC410, lino on floor	Cushion floor (paper)	5 m^2	Easy	Good - damage to edge	Covered by vinyl	Chrysotile	4	Sampled 1 sample	4	11
Corridor, BC411, electrical switch box	Woven cloth	Possibly 4 items	Medium	Medium	Unsealed	Chrysotile	5	Strongly presumed	8	14 remove during next campaign
Plant room 2, BC416, lift motor	Brake shoes	2 items	Difficult	Medium	Unsealed	Chrysotile	6	Strongly presumed	4	10 'H' Vac dust
Plant room 2, BC416, pipe lagging	Pipe insulation	24 linear metres	Easy	Good	Sealed and labelled	Crocidolite Amosite Chrysotile	7	Sampled 6 samples	8	14 remove during next campaign
Plant room 2, BC416, wall panels	Asbestos panels (AIB)	43 m^2	Easy	Good	1 face sealed and labelled	Chrysotile	8	Sampled 4 samples	5	14 monitor weekly

Material scores above 10 have high potential to release fibres

Table 1 Summary of survey results and format for asbestos register

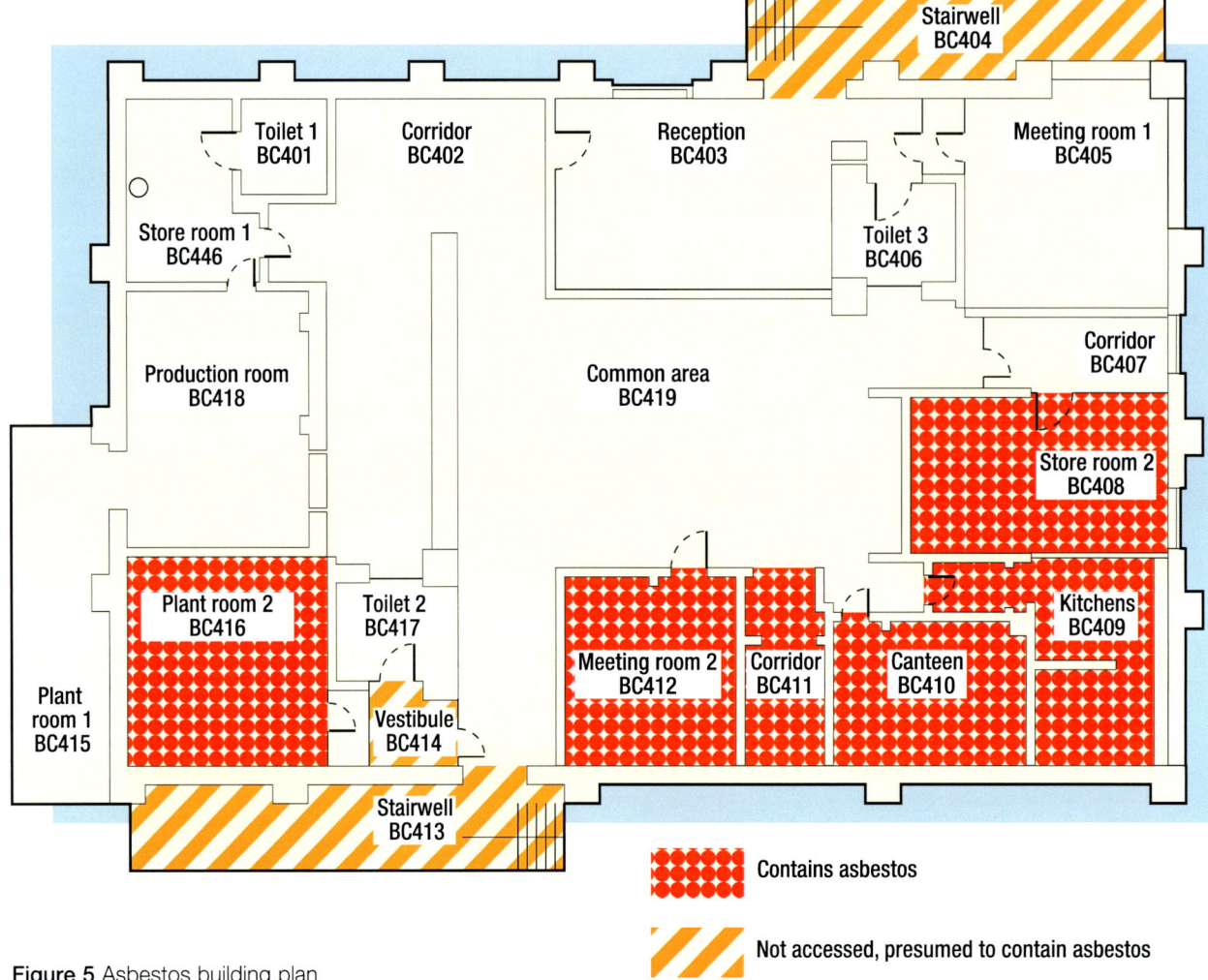

Figure 5 Asbestos building plan

Conclusions and actions

142 The conclusions section should summarise the rooms where asbestos is present and the products/items which contain asbestos (ie it should be an 'easy guide' for the client/dutyholder). It should also contain a list of any actions identified in the material assessment (or the priority assessment if included) and indicate their urgency, eg immediate, middle/longer term.

Bulk analysis results

143 The survey report should also include the certificate of analysis showing the results of the samples taken. This data can be listed in an appendix with the following information:

- the name and address of the laboratory carrying out the bulk identification;
- a reference to the method used;
- the laboratory's current UKAS accreditation for bulk asbestos analysis/sampling and accreditation number;
- a table or appendix summarising the results of the bulk analysis, including asbestos found or not found and types identified, by sample identifier;
- dates the bulk analysis was carried out and reported by the laboratory; and
- the names and signatures of the analyst and any countersigning person.

144 Photographs can be very informative to the client and should be included in the report. Photographs can show the material sampled, its condition and location and its surrounding environment. Photographs provide a context for the sample and can assist the client in managing asbestos for example by providing a benchmark for the comparison of condition over time. Photographs can also be used to identify the actual sampling points. However, it is important not to dominate the report with photographs. It is not necessary to have a single page per photograph. It is also worth noting that much of the detail of photographs can be obscured in photocopied reports.

7 Dutyholder's use of survey information

145 The survey report needs to meet the requirements of the client and comply with the tender/contractual obligations. The report should be fit for purpose and the client should check that this is the case. Therefore, the client should examine the report and carry out a number of checks to make sure that the survey has been adequate and that the report is suitable and accurate. These checks are set out in Box 11.

Box 11: What the client/dutyholder should do to check the accuracy of the survey report

- Check the report against the original tender.
- Check for unagreed caveats or disclaimers.
- Check that the survey is as requested: Management or refurbishment/demolition (or a combination).
- Check diagrams and plans are clear and accurate.
- Check all rooms and areas have been accessed.
- Check sufficient samples have been taken (usually 1–2 per area/room) and that sample numbers are not disproportionate (eg dominated by one ACM type).
- Check sample numbers reflect variations in the same ACMs, eg different ceiling tiles in the same room.
- Check for any obvious discrepancies and inconsistencies.

146 The information in the survey report should be used to form the asbestos register which is a key component of the management plan. The survey report itself will generally **not** be the asbestos register. The asbestos register will be a simpler document and will not contain most of the information in the survey report, eg the bulk analysis results or survey site information. However, the results in the survey report should be presented in a manner or format that can be directly lifted or employed to form the asbestos register. Table 1 shows a sample asbestos register.

147 The asbestos register is a living document which must always contain current information on the presence and condition of asbestos. The register will need regular updating. The dutyholder should make:

- deletions to the register when asbestos is removed;
- additions to the register when new areas are sampled;
- changes to the register if the condition of ACMs has altered on rechecking.

148 The asbestos register can be a paper or electronic document/database. It must be up to date and accessible, irrespective of the form. Hard copies are useful for ease of interrogation and may suit dutyholders with smaller and simpler premises. Electronic versions are often more convenient for dutyholders with large numbers of properties. Electronic documents can be readily accessible and easily updated. Electronic documents can also be used to record all the remedial work carried out and to prompt the relevant person, eg the building manager, to carry out and record any further inspections required. Some databases can also link digital picture images of a sample and CAD plans. Printed versions of an electronic document will usually be needed for contractors and others. The distribution and circulation of such documents will need to be carefully managed to ensure only current versions are referenced.

149 The asbestos register must be available to those who plan or initiate maintenance and related work, so it can be consulted before the work is authorised. The asbestos register should be consulted for all work which may disturb the fabric of the building or involve the building services. It includes simple and short duration work such as single hole-drilling or attaching items to walls. The register must be kept in convenient locations (eg in the estates or building manager's or appointed person's office) and be easily accessible. Information on the presence of asbestos should be passed to contractors as early as possible (eg at the tender or engagement stage of work) so that appropriate precautions, procedures and controls can be employed. Supplying the information or the register to a contractor when they arrive on site to do the work is unlikely to provide sufficient time for the correct control arrangements to be put into place. The dutyholder should ensure that the contractor acknowledges the presence of ACMs and will implement controls. The register should also be available to contractors and employees on request.

Appendix 1: Refurbishment and demolition surveys

1 Refurbishment and demolition surveys are technically more challenging than management surveys, as their purpose is to identify all ACMs within a particular building area or within the whole premises, so they can be removed. Many buildings have been individually designed with their own layout and materials. There may have been numerous refurbishments and modifications over the years, with many changes and alterations to the building structure and appearance, eg 'false' floors, ceilings and walls, concealed and hidden areas and surface treatments. Building drawings may not have been updated. Management surveys will not have accessed structural locations (eg behind concrete or between floors and walls such as cavity walls). The level of competency and knowledge needed for refurbishment and demolition surveys (eg on construction building techniques) is much greater than for management surveys and the intrusive nature presents more health and safety hazards.

2 This appendix provides additional practical guidance to help complete refurbishment and demolition surveys. It provides details of specific areas which should be inspected in a refurbishment or demolition survey (but note that the list is not exhaustive). The guidance also tries to describe these areas and illustrate how to collect samples. These items are **in addition** to those normally found in a management survey.

3 In many instances, access will be required into the fabric of the building and various items such as brickwork, timber, boards and panels etc will have to be removed or broken into. In these circumstances, it may be helpful to seek professional help on such activities/work from a joiner, builder, maintenance worker, engineer or other appropriate person. In some situations, where concrete is to be sampled or brickwork removed, advice may have to be sought from a competent person, such as a structural engineer.

Textured coatings

4 The more explicit guidance in *Work with materials containing asbestos* on removal of ACMs from buildings before demolition applies directly to textured coatings. ACMs should only be removed before demolition if it is reasonably practicable. In many cases, it will **not** be reasonably practicable to remove textured coatings before demolition as the removal is resource-intensive/time-consuming and involves other risks, eg where textured coatings are attached directly to substrates such as concrete or lath. However, where textured coatings are attached onto materials which can be essentially removed intact or whole, eg plasterboard sheets, then removal may be reasonably practicable, eg by removing whole sheets intact. The survey should identify the nature of the substrate and whether textured coating removal will be required. Textured coating removal will be necessary where refurbishment is taking place.

Areas to be examined

'No access' areas from previous survey
5 All 'no access' areas on previous surveys (if available) must be accessed with suitable access equipment and procedures.

Suspended ceilings
6 Suspended ceilings not previously accessed (eg AIB tiles screwed to wooden battens) must be entered, by means of an enclosure and airlock system constructed by a licensed asbestos-removal contractor. This work is likely to exceed the short-term exemption under the Approved Code of Practice *Work with materials containing asbestos* and will therefore be notifiable to the relevant enforcing authority.

7 If the work is deemed to be short duration and non-licensed, ie it meets the short-term exemption (as well as the other criteria) under the Approved Code of Practice *Work with materials containing asbestos* (ie total time less than two hours with no one person working for more than one hour), then the work may be done by a competent non-licensed contractor using appropriate control measures (eg *Asbestos essentials*).[17] However, a licensed contractor is strongly recommended.

8 Suspended ceiling voids are often very cluttered and full of ventilation ducting, pipework and cables. If several entry points are required or more than one void is to be inspected, the work will probably be notifiable and require a licensed asbestos-removal contractor.

9 The ceiling void may contain asbestos debris and other ACMs such as sprayed coating, pipe insulation, older ceilings above the latest one, damaged fire breaks etc. All ACMs in the void will need to be located.

Partition walls (plasterboard/AIB sandwich)

10 Walls may not be uniform and may have undergone partial replacement. All sections of a partition wall will need to be examined, unless documentary evidence confirms that they were erected at a time when ACMs would not have been used or the original specification confirms that ACMs were not to be used. Visual inspection will not be enough on its own. If the evidence is that the walls were erected at a specific time and that no refurbishment or alteration has taken place, then an appropriate proportion of the sections should be examined.

11 The joints between partition panels may contain asbestos rope fire seals. The rope may only be apparent when the outer trim (eg aluminium) is removed.

Cavity walls

12 Although loose asbestos was not known to be used as a cavity insulation material, wall cavities should be inspected with an endoscope to check for the presence of any asbestos materials or debris such as AIB. Entry points should be agreed with a competent person, eg a builder, joiner or structural engineer.

13 Walls should also be examined thoroughly where insulated heating pipes pass through brick or breeze block walls. Check for insulation or residues within the wall cavity itself.

Apertures (doors, windows etc)

14 Cavity closers (usually AC) are sometimes found around air bricks, windows etc. All apertures should be considered and examined thoroughly.

15 Window frames commonly had AIB packers or spacers where the window frame was attached to the brick wall. Asbestos rope seals as fire breaks are also found.

16 Door frames (particularly around fire doors) should be inspected for AIB packers where the frame is fitted into the doorway. The architraves will need to be removed.

Floors

17 Carpets and tiles must be lifted. The floor tile adhesives also frequently contained asbestos.

18 Floor ducts or trenches must be accessed and inspected for shuttering, services, pipe insulation, fire stops, debris etc. The inspection includes the duct cover itself, which may have AC or AIB shuttering. The full length of each duct will need to be inspected, unless it is clear that asbestos pipe insulation is present throughout, when the entire run can be treated as containing asbestos.

19 Floor boards must be lifted to examine the void below. Sufficient boards must be lifted to ensure that the whole floor void is examined for loose asbestos, AIB debris, packers, fire protection, electric cables etc. It may be necessary to inspect the ends of the joists for AIB packing.

20 Slab (poured concrete) floors are known to contain AIB or AC which was used as an expansion joint or shuttering below the surface. These may only be found by drilling core samples through the slab. This will need specialist advice on the structural considerations and on the equipment needed to carry out this type of investigation.

21 AC sleeves were used where cables or pipes run through a slab floor, although these should be visible at the surface.

Ducts

22 Service risers, including fire stops between floors, if not investigated under a previous survey, must be inspected. Lift shafts must be inspected, including the pit at the bottom of the shaft. Ventilation shafts or ducts have been seen with asbestos acoustic attenuators and with debris from assorted ACMs. Ventilation trunking should also be examined.

Cladding

23 Columns or stanchions may have been originally provided with fire protection from AIB or sprayed coating. It may be concealed by over-cladding with a non-asbestos board (Supalux or Masterboard), wood or metal sheet. Inspect all columns.

24 External cladding of tiles or slates (which may or may not be asbestos) will usually conceal a moisture membrane based on a bituminous ACM and possibly AIB panels.

Debris in boiler room areas

25 This should have been investigated during a previous survey (eg old Type 2 survey) but it may be necessary to look closely at where pipes pass through walls, or in sumps and gulleys, behind and underneath tanks and other plant. In particular, the walls and floors should be inspected for insulation debris, which may have been painted over. All plant and electrical equipment must be investigated (while certified in a safe condition) (but see note on consumer electrical equipment in paragraph 96).

26 It may not be possible to locate some or all of the debris until the plant (tanks or boilers) has been removed. It will be necessary to remove the plant under controlled conditions with an appropriate plan of work. Cast iron sectional boilers with asbestos between the sections (or as a plinth under the boiler) will need to be disassembled under controlled conditions.

Debris underneath non-asbestos reinsulation

27 If the desk-top study reveals that asbestos insulation has been stripped and replaced, a proportion of the new insulation must be removed to examine the extent of any asbestos debris on the pipes, bolt-heads and flanges. If any of the pipes are shown to have frequent occurrences of asbestos debris, then it is likely that the pipes will have to be removed as ACMs.

Roof voids

28 Where Rockwool or vermiculite loft insulation is present in a roof void, the areas underneath it should be inspected, particularly if there is evidence of other ACMs, such as AIB as fire breaks etc.

29 Loose asbestos is very occasionally found as loft insulation in houses around old asbestos factories or dockyards.

Previously demolished areas: From the desk-top study

30 The desk-top study should be used to investigate whether any previous structures (including underground structures) remain or may have released asbestos debris into the soil.

31 Whether the desk-top study information is available or not, the site must be inspected visually to identify obvious signs of demolition works and associated surface asbestos debris. It may be necessary to treat the external area as a contaminated site for investigation purposes, in which case, trenches and pits may need to be excavated to establish the extent of the debris.

32 The desk-top study will need to include reference to old plans from historical archives, for example.

Overspray debris from sprayed coatings

33 If a sprayed asbestos coating is present or known to have been present at some time in the past, the area must be inspected carefully for the presence of debris and to establish the extent and location of any overspray.

Use of AIB as packing and shuttering

34 Depending on the age of the building, surveyors need to be vigilant in buildings constructed in the 1960s and 1970s for the use of AIB as packing and shuttering. This was frequently used simply as a convenient piece of board and not because of the need for fire protection etc. Some of these applications should have been found under a previous survey (eg old Type 2 survey).

Damp-proof course (DPC)

35 Any DPC with asbestos should have been detected in a previous survey (eg old Type 2 survey). It will not normally be necessary to remove this during demolition.

36 It must be emphasised that this is a list of common and/or frequently found locations which must be examined. It must not be regarded as exhaustive or exclusive, and each type of structure must be thoroughly examined on its own merits.

Appendix 2: ACMs in buildings listed in order of ease of fibre release

Asbestos product	Location/use	Asbestos and type/date last used	Ease of fibre release and product names
Loose insulation			
Bulk loose fill, bulk fibre-filled mattresses, quilts and blankets. Also 'jiffy bag' type products used for sound insulation.	Bulk loose fill insulation is now rarely found but may be encountered unexpectedly, eg DIY loft insulation and fire-stop packing around cables between floors. Mattresses and quilts used for thermal insulation of industrial boilers were filled with loose asbestos. Paper bags/sacks were also loose-filled and used for sound insulation under floors and in walls.	Usually pure asbestos except for lining/bag. Mattresses and quilts usually contain crocidolite or chrysotile. Acoustic insulation may contain crocidolite or chrysotile.	Loose asbestos may readily become airborne if disturbed. If dry, these materials can give rise to high exposures. Covers may deteriorate or be easily damaged by repair work or accidental contact.
Sprayed coatings			
Dry applied, wet applied and trowelled finish.	Thermal and anti-condensation insulation on underside of roofs and sometimes sides of industrial buildings and warehouses. Acoustic insulation in theatres, halls etc. Fire protection on steel and reinforced concrete beams/columns and on underside of floors. Overspray of target areas is common.	Sprayed coatings usually contain 55%–85% asbestos with a Portland cement binder. Crocidolite was the major type until 1962. Mixture of types including crocidolite until mid-1971. Asbestos spray applications were used up to 1974.	The surface hardness, texture and ease of fibre release will vary significantly depending on a number of factors. Sprays have a high potential for fibre release if unsealed, particularly if knocked or the surface is abraded or delaminates from the underlying surface. Dust released may then accumulate on false ceilings, wiring and ventilation systems. 'Limpet' (also used for non-asbestos sprays).
Thermal insulation			
Hand-applied thermal lagging, pipe and boiler lagging, pre-formed pipe sections, slabs, blocks. Also tape, rope, corrugated paper, quilts, felts, and blankets.	Thermal insulation of pipes, boilers, pressure vessels, calorifiers etc.	All types of asbestos have been used. Crocidolite used in lagging until 1970. Amosite was phased out by the manufacturers during the 1970s. Content varies 6–85%. Various ad hoc mixtures were hand-applied on joints ▶	The ease of fibre release often depends on the type of lagging used and the surface treatment. Often it will be encapsulated with calico and painted (eg PVA, EVA, latex, bitumen or proprietary polymer emulsions or ▶

Asbestos product	Location/use	Asbestos and type/date last used	Ease of fibre release and product names
Thermal insulation (continued)			
		and bends and pipe runs. Pre-formed sections were widely used, eg '85% magnesia' contained 15% amosite, 'Caposil' calcium silicate slabs and blocks contained 8–30% amosite while 'Caposite' sections contained ~ 85% amosite. Blankets, felts, papers, tapes and ropes were usually ~100% chrysotile.	PVC, neoprene solutions), eg 'Decadex' finish is a proprietary polymer emulsion. A harder chemical-/weather-resistant finish is known as 'Bulldog'.
Asbestos boards			
'Millboard'.	'Millboard' was used for general heat insulation and fire protection. Also used for insulation of electrical equipment and plant.	Crocidolite was used in some millboard manufacture between 1896 and 1965; usually chrysotile. Millboards may contain 37–97% asbestos, with a matrix of clay and starch.	Asbestos 'Millboard' has a high asbestos content and low density so is quite easy to break and the surface is subject to abrasion and wear.
Insulating board.	Used for fire protection, thermal and acoustic insulation, resistance to moisture movement and general building board. Found in service ducts, firebreaks, infill panels, partitions and ceilings (including ceiling tiles), roof underlay, wall linings, soffits, external canopies and porch linings.	Crocidolite used for some boards up to 1965, amosite up to 1980, when manufacture ceased. Usually 15–25% amosite or a mixture of amosite and chrysotile in calcium silicate. Older boards and some marine boards contain up to 40% asbestos.	AIB can be readily broken, giving significant fibre release. Also significant surface release is possible by abrasion, but surface is usually painted or plastered. Sawing and drilling will also give significant releases. 'Asbestolux', 'Turnasbestos', 'LDR', 'asbestos wallboard', 'insulation board'. Marine boards known as 'Marinite' or 'Shipboard'.
Insulating board in cores and linings of composite products.	Found in fire doors, cladding infill panels, domestic boiler casings, partition and ceiling panels, oven linings and suspended floor systems. Used as thermal insulation and sometimes as acoustic attenuators.	Crocidolite used for some boards up to 1965, amosite up to 1980, when manufacture ceased. 16–40% amosite or a mixture of amosite and chrysotile.	Can be broken by impact; significant surface release possible by abrasion, but usually painted or plastered. Sawing and drilling will also give significant releases. 'Asbestolux'. Caposil.

Asbestos product	Location/use	Asbestos and type/date last used	Ease of fibre release and product names
Paper, felt and cardboard			
	Used for electrical/heat insulation of electrical equipment. Also used in some air-conditioning systems as insulation and acoustic lining. Asbestos paper has also been used to reinforce bitumen and other products and as a facing/lining to flooring products, combustible boards, flame-resistant laminate. Corrugated cardboard has been used for duct and pipe insulation.	Asbestos paper can contain ~100% chrysotile asbestos but may be incorporated as a lining, facing or reinforcement for other products, eg roofing felt and damp-proof courses, steel composite wall cladding and roofing (see asbestos bitumen products below), vinyl flooring. Asbestos paper is also sometimes found under MMMF insulation on steam pipes.	Paper materials, if not encapsulated/combined within vinyl, bitumen, or bonded in some way, can easily be damaged and release fibres when subject to abrasion or wear (eg worn flooring surface with paper backing). Asbestos paper, asbestos felt, 'Novilon' flooring, Durasteel laminates, vinyl asbestos tile, roofing felt and damp-proof course etc. 'Pax felt'. 'Viceroy' (foil-coated paper). 'Serval'.
Textiles			
Ropes and yarns.	Used as lagging on pipes (see above), jointing and packing materials and as heat/fire-resistant boiler, oven and flue sealing. Caulking in brickwork. Plaited asbestos tubing in electric cable.	Crocidolite and chrysotile were widely used due to length and flexibility of fibres. Other types of asbestos have occasionally been used in the past. Chrysotile alone since at least 1970. Asbestos content approaching 100% unless combined with other fibres.	Weaving reduces fibre release from products, but abrading or cutting the materials will release fibres, likely to degrade if exposed, becoming more friable with age. If used with caulking, fibres will be encapsulated and less likely to be released.
Cloth.	Thermal insulation and lagging (see above), including fire-resisting blankets, mattresses, protective curtains, gloves aprons and overalls. Curtains, gloves etc were sometimes aluminised to reflect heat.	All types of asbestos were used. Since the mid-1960s the vast majority have been chrysotile. Asbestos content approaching 100%.	Fibres may be released if material is abraded.
Gaskets and washers.	Used widely in domestic and industrial plant and pipe systems ranging from hot water boilers to industrial power and chemical plant.	Variable but usually around 90% asbestos, crocidolite used for acid resistance and chrysotile for chlor-alkali. Some gasket materials continued to be used after asbestos prohibition in 1999 (through exemption).	May be dry and damage easily when removed. Mainly a problem for maintenance workers. 'Klingerit', 'Lion jointing', 'Permanite', 'CAF' – compressed asbestos fibre or 'It' in German gaskets.
Strings.	Used for sealing hot water radiators.	Strings have asbestos content approaching 100%.	

Asbestos product	Location/use	Asbestos and type/date last used	Ease of fibre release and product names
Friction products			
Resin-based materials.	Transport, machinery and lifts, used for brakes and clutch plates.	30–70% chrysotile asbestos bound in phenolic resins. Used up to November 1999.	Normal handling will produce low emissions. Minor emissions when braking. Dust may build up with friction debris. Grinding brake and clutch components to fit and brushing or blowing clean can produce significant peak airborne levels.
Drive belts/conveyor belts.	Engines, conveyors.	Chrysotile textiles encapsulated in rubber.	Low friability, except when worn to expose textile.
Cement products			
Profiled sheets.	Roofing, wall cladding. Permanent shuttering, cooling tower elements.	10–15% asbestos (some flexible sheets contain a proportion of cellulose). Crocidolite (1950–1969) and amosite (1945–1980) have been used in the manufacture of asbestos cement, although chrysotile (used until November 1999) is by far the most common type found.	Likely to release increasing levels of fibres if abraded, hand sawn or worked on with power tools. Exposed surfaces and acid conditions will remove cement matrix and concentrate unbound fibres on surface and sheet laps. Cleaning asbestos-containing roofs may also release fibres.
			Asbestos cement, Trafford tile, 'Bigsix', 'Doublesix', 'Supersix', 'Twin twelve', 'Combined sheet', 'Glen six', '3' and 6' corrugated', 'Fort', 'Monad', 'Troughsec', 'Major tile and Canada tile', 'Panel sheet', 'Cavity decking'.
Semi-compressed flat sheet and partition board.	Partitioning in farm buildings and infill panels for housing, shuttering in industrial buildings, decorative panels for facings, bath panels, soffits, linings to walls and ceilings, portable buildings, propagation beds in horticulture, domestic structural uses, fire surrounds, composite panels for fire protection, weather boarding.	As for profiled sheets. Also 10–25% chrysotile and some amosite for asbestos wood used for fire doors etc. Composite panels contained ~ 4% chrysotile or crocidolite.	Release as for profiled sheets. Flat building sheets, partition board, 'Poilite'.

Asbestos product	Location/use	Asbestos and type/date last used	Ease of fibre release and product names
Cement products (continued)			
Fully compressed flat sheet used for tiles, slates, board.	As above, but where stronger materials are required, and as slates, board cladding, decking and roof slates (eg roller-skating rinks, laboratory worktops). Higher asbestos content sheets produced for industrial applications as a high grade arc and heat-resistant material.	As for profiled sheets. Up to 50% chrysotile.	Release as for profiled sheets. Asbestos-containing roofing slate (eg 'Eternit', 'Turners', 'Speakers'), 'Everite', 'Turnall', 'Diamond AC', 'JM slate', 'Glasal AC', 'Emalie, Eflex', 'Colourglaze', 'Thrutone', 'Weatherall'. 'Sindanyo'.
Pre-formed moulded products and extruded products.	Cable troughs and conduits. Cisterns and tanks. Drains and sewer pressure pipes. Fencing. Flue pipes. Rainwater goods. Roofing components (fascias, soffits etc). Ventilators and ducts. Weather boarding. Window sills and boxes, bath panels, draining boards, extraction hoods, copings, promenade tiles etc.	As for profiled sheets.	Release as for profiled sheets. 'Everite', 'Turnall', 'Promenade tiles'.
Other encapsulated materials			
Textured coatings.	Decorative/flexible coatings on walls and ceilings.	3–5% chrysotile asbestos. Chrysotile added up to 1984 but old stock may have been used for several more years. Non-asbestos versions were available from the mid-1970s.	Generally fibres are well contained in the matrix but may be released when old coating is sanded down or scraped off. 'Artex', 'Wondertex', 'Suretex', 'Newtex', 'Pebblecoat', 'Marblecoat'.
Bitumen products.	Roofing felts and shingles, semi-rigid asbestos bitumen roofing. Gutter linings and flashings. Bitumen damp-proof courses (dpc). Asbestos/bitumen coatings on metals (eg car body underseals). Bitumen mastics and adhesives (used for floor tiles and wall coverings).	Chrysotile fibre or asbestos paper (approximately 100% asbestos) in bitumen matrix, usually 8% chrysotile. Used up to 1992. Adhesives may contain up to a few per cent chrysotile asbestos. Used up to 1992.	Fibre release unlikely during normal use. Roofing felts, dpc and bitumen-based sealants must not be burnt after removal. See felts and papers.

Asbestos product	Location/use	Asbestos and type/date last used	Ease of fibre release and product names
Other encapsulated materials (continued)			
Flooring.	Thermoplastic floor tiles.	Up to 25% asbestos.	Fibre release is unlikely to be a hazard under normal services conditions. Fibre may be released when material is cut, and there may be substantial release where flooring residue, particularly paper backing, is power-sanded. 'Novilon', 'Serval asbestos'. Very hard, fibre release unlikely.
	PVC vinyl floor tiles and unbacked PVC flooring.	Normally 7% chrysotile.	
	Asbestos paper-backed PVC floors.	Paper backing approximately 100% chrysotile asbestos. Used up to 1992.	
	Magnesium oxychloride flooring used in WCs, staircases and industrial flooring.	About 2% asbestos.	
Reinforced PVC.	Panels and cladding.	1–10% chrysotile asbestos.	Fibre release is unlikely.
Reinforced plastic and resin composites.	Used for toilet cisterns, seats, banisters, window seals, lab bench tops.	Plastics usually contain 1–10% chrysotile asbestos. Some amphiboles were used to give improved acid resistance, eg car batteries. Resins were reinforced with woven chrysotile cloth, usually contain 20–50% asbestos.	Fibres unlikely to be released, limited emissions during cutting. 'Siluminite', 'Feroasbestos'.

Appendix 3: What ACMs look like and where to find them

1 This appendix gives examples of the main types, locations and uses of ACMs in premises, to help people recognise materials which may contain asbestos. This is only a small selection of the range of ACMs used, but should cover many of the main uses of asbestos in premises.

Loose asbestos insulation

2 Some fire doors contained loose asbestos insulation sandwiched between the wooden or metal facings to give them the appropriate fire rating. Loose asbestos was also packed around electrical cables, sometimes using chicken wire to contain it. Mattresses containing loose asbestos were widely manufactured for thermal insulation. Acoustic insulation has been provided between floors by the use of loose asbestos in paper bags, and in some areas near asbestos works it is not unknown for loose asbestos to have been used as a readily available form of loft insulation.

Figure 6 Loose asbestos used as loft insulation

Sprayed asbestos coatings

3 These are normally homogeneous coatings sprayed or trowelled onto reinforced concrete or steel columns or beams as fireproofing. Sprays were also commonly used on the underside of ceilings for fireproofing and sound and thermal insulation in many high-rise premises. Warehouses and factories commonly had sprayed asbestos applied to walls, ceilings and metal support structures for fireproofing and thermal/anti-condensation insulation purposes. In some larger spaces, sprays were also applied to walls and ceilings for acoustic and decorative purposes (theatres, cinemas, studios, halls etc). The depth of the spray depended on the fire rating and substrate, and may vary from 10 to 150 mm thick. The dry sprayed coatings may have a candyfloss appearance if left untamped (rarely found in the UK). The wet sprayed/trowelled coatings are usually denser, and those with higher proportions of Portland cement that have been well tamped can be quite hard. Surfaces may be sealed with an elasticised paint or proprietary encapsulant, sometimes reinforced with calico or man-made fibre mesh, or left completely unsealed. Spray coatings are vulnerable to accidental damage and also to delamination due to water leakage releasing debris onto the floor and other horizontal surfaces. Overspray onto areas and recesses surrounding the object that was being coated is common. Spray coatings may have deteriorated significantly since installation and must be treated with caution.

Figure 7 Sprayed limpet on car park roof

Figure 8 Sprayed coating on building passageway

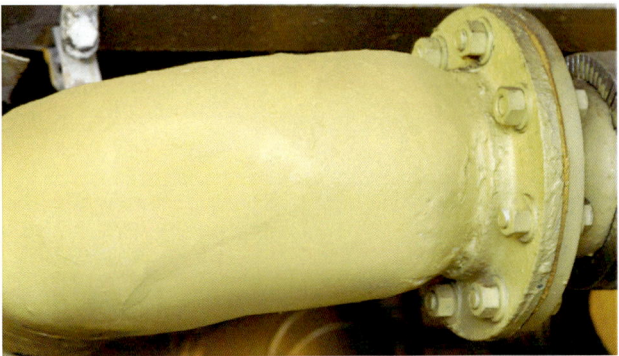

Figure 10 Pipe insulation with coating

Figure 9 Lagged pipe in wall cavity

Figure 11 Amosite lagging in very poor condition on steam pipe

Thermal insulation

4 Asbestos was widely used to insulate pipes, boilers and heat exchangers. There are several types and forms of insulation, often with multi-layer construction. Pre-formed sections of asbestos insulation were made to fit the diameter of the pipe. These would be strapped on and calico-wrapped and sometimes painted (eg 'Decadex' finish), or sealed with a hard plaster (often asbestos-containing) to protect against knocks and abrasion. Other types of asbestos-containing felts, blankets, tapes, ropes and corrugated papers were also used. For bends, joins, small sections of pipe and repairs, an asbestos-containing plaster was wet-mixed on-site and hand-applied to the areas. Larger installations were also insulated with asbestos-containing plaster which was marketed as 'plastic', but various local names were used for this hand-applied insulation (eg 'muck'). Larger thicknesses of insulation would use pre-formed blocks (eg 'Caposil') wired in place, and then various other coatings or layers applied, depending on the insulation required. Very hard-wearing coatings were known as 'Bulldog' finishes and may contain metal sheets and/or chicken wire reinforcement beneath a hard plaster finish. External pipes may also be clad with sheet metal or painted with bitumen for additional weatherproofing. Installers often used whatever materials were available to hand or in stock, so it is very common to find variations on the same pipe or boiler. Pay particular attention to bends and valves, or where it is evident that repairs have been made.

Figure 12 Lagging on large petrochemical vessel

Figure 13 Amosite lagged boiler with hard plastic coating

Millboard

5 Millboard was used when a low-cost, relatively soft low-density board with modest mechanical properties but with good fire, insulation, thermal and electrical properties could be specified. Generally found in industrial premises, but has been used as exterior lining to ventilation ducts and was commonly used inside fire doors.

Figure 14 Asbestos panels inside fire door

Figure 15 AIB panel in porch

Asbestos insulating board (AIB)

6 Widely used in premises for internal partition walls and linings and for fire protection, acoustic and thermal insulation. Suspended ceiling tiles were often made from AIB. Insulating boards come in a range of densities and can be subject to damage by the use of moderate force (eg kicking). There may be variations due to later construction of partition walls as part of a redevelopment or refurbishment. All kinds of combinations are found and surveyors must be alert to all possibilities. Areas around lift shafts, stairwells and service risers in multi-storey buildings were commonly lined or faced with AIB or composites. Similarly, areas around gas fires and central heating boilers were also constructed from AIB. Fire doors were also faced with AIB to achieve the appropriate fire rating. AIB is usually found inside premises, but weather-protected exterior areas, such as porches and soffits, may contain AIB.

Figure 16 AIB exterior painted panel

Figure 17 AIB ceiling tiles

Figure 18 AIB exterior panel on open walkway

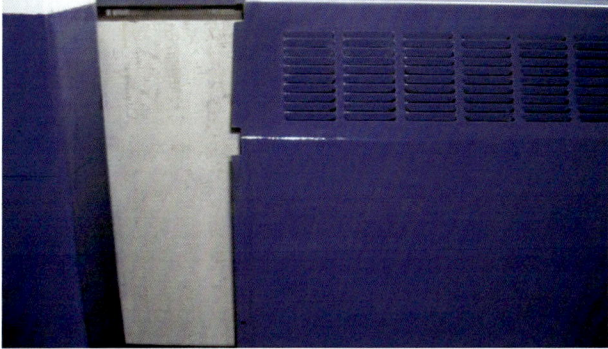

Figure 19 AIB behind school radiator unit

Figure 20 AIB panel between radiator and window

Figure 21 AIB column fire protection in CLASP building

Figure 22 AIB offcuts used as packers around column in CLASP building

AIB in composite materials

7 AIB was used in composite materials and may be sandwiched between or surfaced with non-asbestos products such as strawboard, plywood, metal mesh, sheet metal and plasterboard.

Figure 23 AIB fascia panels with wood effect finish

Asbestos papers, felts and cardboard

8 Air-conditioning trunking may be insulated internally with 'Paxfelt' or externally with other asbestos-containing felt, cardboard and paper for acoustic and heat insulation. Asbestos papers were widely used to line the surfaces of other boards, ceiling tiles and sheet materials.

Figure 24 Chrysotile paper on fibreboard boxing

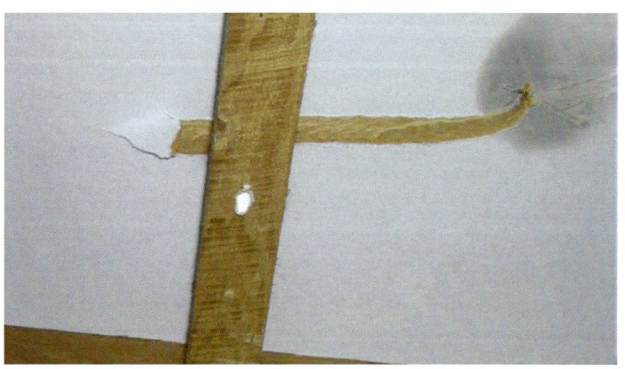

Figure 25 Chrysotile paper on strawboard ceiling panels

Asbestos textiles

9 Asbestos textiles were manufactured for primary heat (eg insulation tapes and ropes) or fire protection uses (eg fire blankets, fire curtains, fire-resistant clothing). Textiles were also used widely as a reinforcing material in friction products/composites.

Figure 26 Asbestos rope seal on drying oven

Figure 27 Amosite asbestos rope packing on riser door frame

Figure 28 Asbestos rope lagging to boiler thermometer

Figure 29 Chrysotile cloth on pipes

Figure 30 Amosite and chrysotile packing on waste pipes

Asbestos gaskets, washers and strings

10 A wide range of asbestos gaskets have been produced and used for sealing pipe and valve joints in industrial plant, but they may also be found in some older domestic boilers etc. Asbestos string was widely used in the past by plumbers for sealing various screw thread joints.

Figure 31 Chrysotile string on skylights

Asbestos cement sheets and tiles used for roofing and cladding

11 Asbestos cement (AC) has been extensively used for roofing and exterior cladding on industrial, public and some domestic premises. Corrugated/profile sheets are commonly found, but flat sheets have also been widely used for exterior and some interior cladding (eg panels below windows and on walls in older prefabricated housing).

Figure 32 Asbestos cement roof and wall cladding

Figure 33 Chrysotile pebble-dash exterior wall coating

Moulded asbestos cement products

12 A wide range of moulded compressed AC products have been used inside premises (eg waste pipes, cold water tanks, flues etc) and outside premises (eg gutters, downpipes, flues, cowls etc). Many other items have been moulded from asbestos cement. Asbestos cement pipes are also used underground (eg from local drainage to regional water supply systems).

Figure 34 Asbestos cement boiler flue

Figure 35 Asbestos cement downpipe

Textured coatings, paints and plasters used for decorative effects

13 These were often manufactured containing up to a few per cent of asbestos. 'Artex', 'Wondertex', 'Suretex', 'Newtex', 'Pebblecoat' and 'Marblecoat' are examples of typical trade products, which usually contained a few per cent of chrysotile asbestos.

Figure 36 Asbestos textured coating on ceiling

Figure 37 Asbestos textured coating on wall

Figure 39 Asbestos-containing vinyl floor tiles

Bitumen products

14 Bitumen-based roofing felts and damp-proof courses have been widely reinforced by the addition of asbestos, usually in the form of chrysotile paper. Bitumen-based wall and floor coverings were also produced. Some mastics used to stick the bitumen products commonly had asbestos added to them to provide flexibility. Other sealants also had asbestos added to improve the performance of the product.

Figure 38 Chrysotile/bitumen coating under metal cladding

Flooring products

15 Polyvinyl chloride (PVC or vinyl) tiles were manufactured with added asbestos to meet a British Standard and often contain a few per cent (5–7%) of very fine chrysotile. Black and brown thermoplastic tiles containing larger amounts and often visible clumps of chrysotile were also produced. Sheet floor coverings were sometimes backed with a thin layer of chrysotile paper (eg 'Novilon', a vinyl flooring, which was more common in Europe). Some underfelts for carpets and linoleum were also manufactured containing asbestos. The mastics which were used to bond the floor covering to the surface could also contain asbestos. Some hard-wearing composite floors (eg magnesium oxychloride) also contain about 2% of mineral fibres which could be asbestos.

Asbestos-reinforced plastic/resin composites and friction products

16 Asbestos-reinforced plastics and resin composite material were used for windowsills, capping for banisters, school and laboratory worktops, toilet cisterns etc. The material is often black and has a high density and scratch resistance. Asbestos textiles were widely used as a reinforcing material in friction products (eg conveyor and fan belts, brake and clutch linings). Older asbestos-containing components may still be in use or present in vehicle repair and maintenance workshops and stores.

Figure 40 Asbestos-reinforced toilet cistern

Figure 41 Asbestos brake linings for quarry vehicles

Figure 42 Asbestos brake lining on output shaft of AC motor

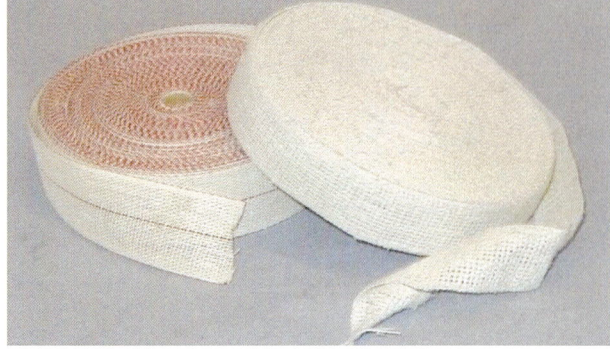

Figure 44 Chrysotile 'scrim' tapes

Metal-asbestos composites

17 Flues for wood-burning stoves were commonly constructed from a metal-asbestos where the asbestos was added as insulation between the inner and outer layers of stainless steel to give a high degree of insulation when passing through floors and on the outside to prevent sudden cooling of the flue gases. 'Durasteel' metal panels were used to provide a strong construction with a certain degree of insulation, by incorporating a layer of asbestos paper.

Domestic appliances and products

19 Many domestic appliances and products contain asbestos insulation materials for thermal or electrical insulation, including ironing boards, hairdryers, oven seals, simmering plates etc. Some older electric fires and storage radiators and old gas fires with catalytic elements or coal or log effect gas fires also contained ACMs.

Figure 43 Metal clad gas flue containing chrysotile lining

Figure 45 'Durasteel' composite steel and asbestos fire door

Wall jointing tapes and fillers

18 Chrysotile textile tapes and webbing were used to reinforce wall joints before plastering. Several types of wall plugs and some wall repair fillers had asbestos added to give additional strength and flexibility. These are very difficult to locate as they are integrated into the plaster finish.

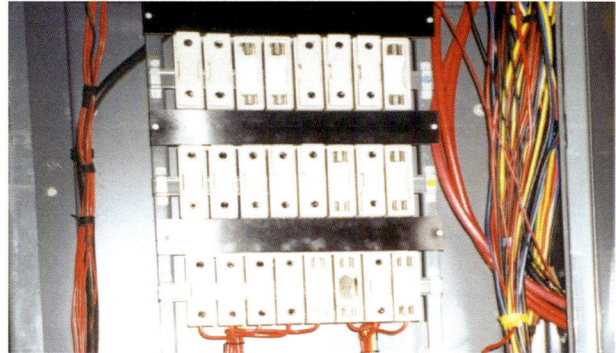

Figure 46 Asbestos tape flash guards in a fuse box

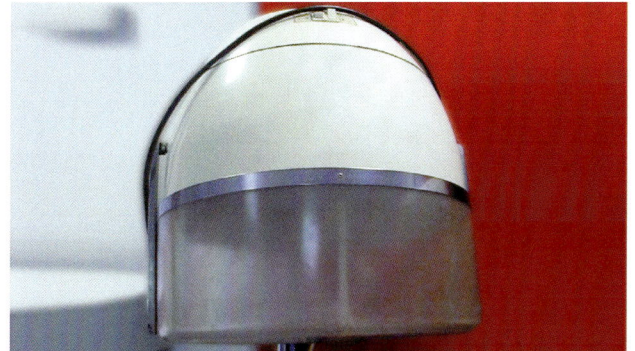

Figure 47 Asbestos-containing commercial hairdryer

Industrial sites, factories and plant

20 Industrial sites (eg refineries, power stations, warehouses and factories) often contain substantial amounts of asbestos. Many of the examples given for spray, thermal insulation and pipe lagging come from industry. Higher-performance ACMs were usually specified to cope with the higher temperatures and pressures prevalent at industrial sites. Some machinery may also incorporate asbestos gaskets and friction products (eg clutches, brake pads, drive belts and conveyor belts). The higher power requirements of industry also saw increased use of asbestos insulation in electrical cables and switchgear.

Figure 48 Asbestos-insulated supply pipes

Figure 49 Electrical switchgear with asbestos flash protection

Figure 50 Open electrical switchboard with asbestos flash guards

Figure 51 Vessel clad in amosite insulation

Asbestos: The survey guide

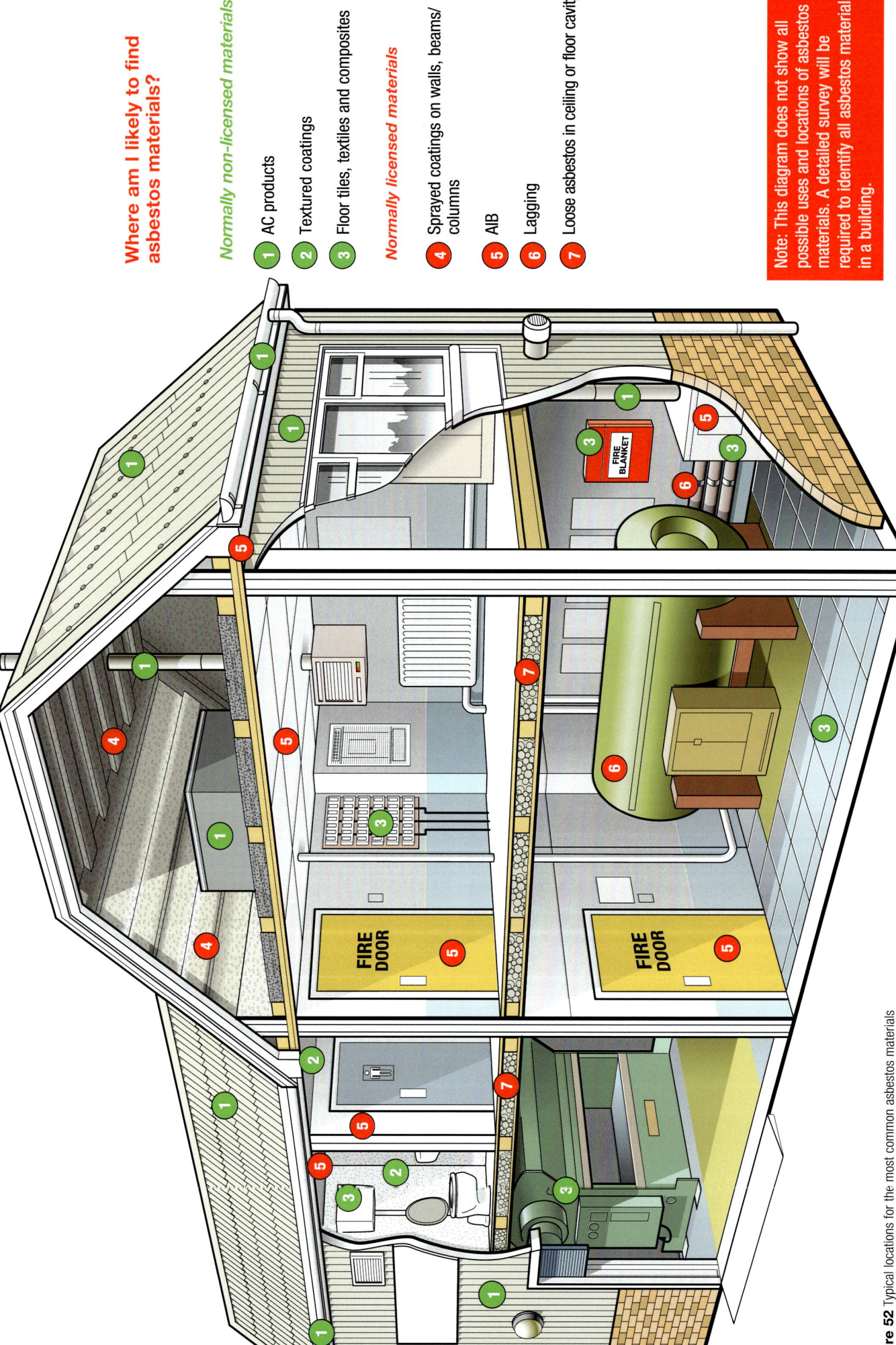

Figure 52 Typical locations for the most common asbestos materials

Appendix 4: Material assessment algorithm

Sample variable	Score	Examples of scores (see notes for more detail)
Product type (or debris from product)	1	Asbestos-reinforced composites (plastics, resins, mastics, roofing felts, vinyl floor tiles, semi-rigid paints or decorative finishes, asbestos cement etc).
	2	AIB, millboards, other low-density insulation boards, asbestos textiles, gaskets, ropes and woven textiles, asbestos paper and felt.
	3	Thermal insulation (eg pipe and boiler lagging), sprayed asbestos, loose asbestos, asbestos mattresses and packing.
Extent of damage/deterioration	0	Good condition: no visible damage.
	1	Low damage: a few scratches or surface marks, broken edges on boards, tiles etc.
	2	Medium damage: significant breakage of materials or several small areas where material has been damaged revealing loose asbestos fibres.
	3	High damage or delamination of materials, sprays and thermal insulation. Visible asbestos debris.
Surface treatment	0	Composite materials containing asbestos: reinforced plastics, resins, vinyl tiles.
	1	Enclosed sprays and lagging, AIB (with exposed face painted or encapsulated) asbestos cement sheets etc.
	2	Unsealed AIB, or encapsulated lagging and sprays.
	3	Unsealed lagging and sprays.
Asbestos type	1	Chrysotile.
	2	Amphibole asbestos excluding crocidolite.
	3	Crocidolite.
Total		

Score	Potential to release asbestos fibres
10 or more	High
7–9	Medium
5–6	Low
4 or less	Very low

Non-asbestos materials have no potential to release asbestos fibres

Appendix 5: Example of a survey and sampling equipment checklist

Survey equipment

- Site plan.
- Log book, organiser, computer.
- Step ladder.
- Camera (film or digital) with flash and preferably with a date and number facility.
- Torch.
- Access keys to rooms and covers.
- Screwdrivers.

PPE for sampling

- Disposable overalls (hooded).
- Disposable overshoes or Wellington boots.
- Disposable gloves.

Bulk sampling equipment

- Pliers.
- Screwdrivers.
- Core samplers or cork borers.
- Aluminium foil or cloth tape.
- Stanley knife with spare blades.
- Hand-spray with diluted PVA or surfactant.
- Sample bags (polythene self-seal bags).
- Sample point labels.
- Type H vacuum.
- Asbestos waste bags of the approved type.
- Warning signs: 'Asbestos sampling: Keep clear'.
- Wet wipes and tissues.
- Polythene sheeting.

RPE

- As per assessment.

Appendix 6: Quality assurance and quality control

1   All organisations providing an asbestos surveying service (including 'sole traders') should have an adequate quality management system, including quality control of survey work. Organisations accredited under UKAS to ISO/IEC 17020 will already have appropriate quality management schemes in place.

2   Non-accredited organisations (including surveyors holding recognised personnel certification) should also implement an effective quality management system to support their work. Quality management systems are set out in ISO/IEC 17020 (or ISO 9001 as a minimum).

3   The following paragraphs outline three of the essential components of a quality management system. Sole traders are not exempt from the need to have such a system and should implement self-checking versions (or other working arrangements) for quality assurance and checking survey reports. It should be possible to engage an independent organisation to conduct an annual audit of completed surveys. All surveying organisations should have written quality management procedures and keep records of their audits and checks.

Quality assurance for site work

4   A proportion of surveys should be 'reinspected' (ie rechecked) while the survey is still in progress. It is recommended that about 5% of all surveys are reinspected. The process of site selection is at the discretion of the organisation, but the system should ensure that the sites selected:

- are representative of the different types of survey that the organisation performs;
- are representative of the different types of premises surveyed;
- cover all the surveyors employed.

5   Inspections or audits of newly qualified or recently employed surveyors should be more frequent until it can be established that they are capable of consistently working to the required standards.

6   In some situations it may not be practical (eg very large surveys) to reinspect the whole site. In these circumstances a representative part of the site should be re-examined.

7   The survey reinspection will involve checking all aspects of the site work using the recorded data, samples and photographs to ensure:

- no ACMs or suspect ACMs were omitted from the recorded data;
- all recorded ACMs and suspect ACMs were valid;
- where suspect ACMs have been 'presumed' or 'strongly presumed', the presumption of asbestos type is valid;
- all identifiers for records, sample numbers and photograph numbers correspond and are unique;
- all areas inspected were correctly and unambiguously identified;
- all 'no access' areas were valid and were correctly and unambiguously identified;
- all material types for ACMs and suspect ACMs were correctly listed;
- all recorded ACMs and suspect ACMs were correctly and uniquely located;
- all quantities of materials and suspect ACMs were correctly assessed;
- the correct assessments have been made and recorded for:
  - asbestos product type;
  - surface treatment;
  - damage;
  - accessibility (vulnerability);
- adequate numbers and sizes of samples were collected, correctly labelled and individually double bagged;
- adequate cleaning has occurred after sampling; and
- sampling sites have been made good in the agreed manner and in accordance with the plan of work.

(Where omissions, deficiencies or errors are identified, there should be arrangements in place to rectify the situation including retraining and supervision of personnel where appropriate.)

Audit of completed surveys

8 There should be an annual audit of the management systems and procedures in place. It would normally be a desk-top audit (as the site may have changed from the original survey, eg undergone refurbishment. However, a full site resurvey may be necessary if, for example, significant anomalies were discovered). The audits should include reviewing:

- report formats, structure and content;
- raw data transposition into report;
- authorisation or approval of report checker, surveyor – their authorisations, training records and qualifications etc;
- contract review – documented records of the client's instructions – ensure that the report meets these in full;
- records and storage of raw data, site logs etc, as well as compiled reports.

Survey reports

9 Every report produced should be checked by an authorised person before being issued to the client. The checks must ensure that the report contents are technically consistent, accurate and complete.

10 In particular, check:

- the client's instructions for the survey and report have been followed;
- all site notes agree with the final report;
- no observed ACMs have been omitted;
- all appendices (eg certificates of analysis) are included as required;
- all titles, reference numbers and descriptions are correct;
- the assessments and recommendations for any remedial work are appropriate; and
- the report summary is included and is a fair statement.

11 **Non-accredited organisations may find it helpful to consult the UKAS document** *Accreditation of bodies surveying for asbestos in premises.*[18]

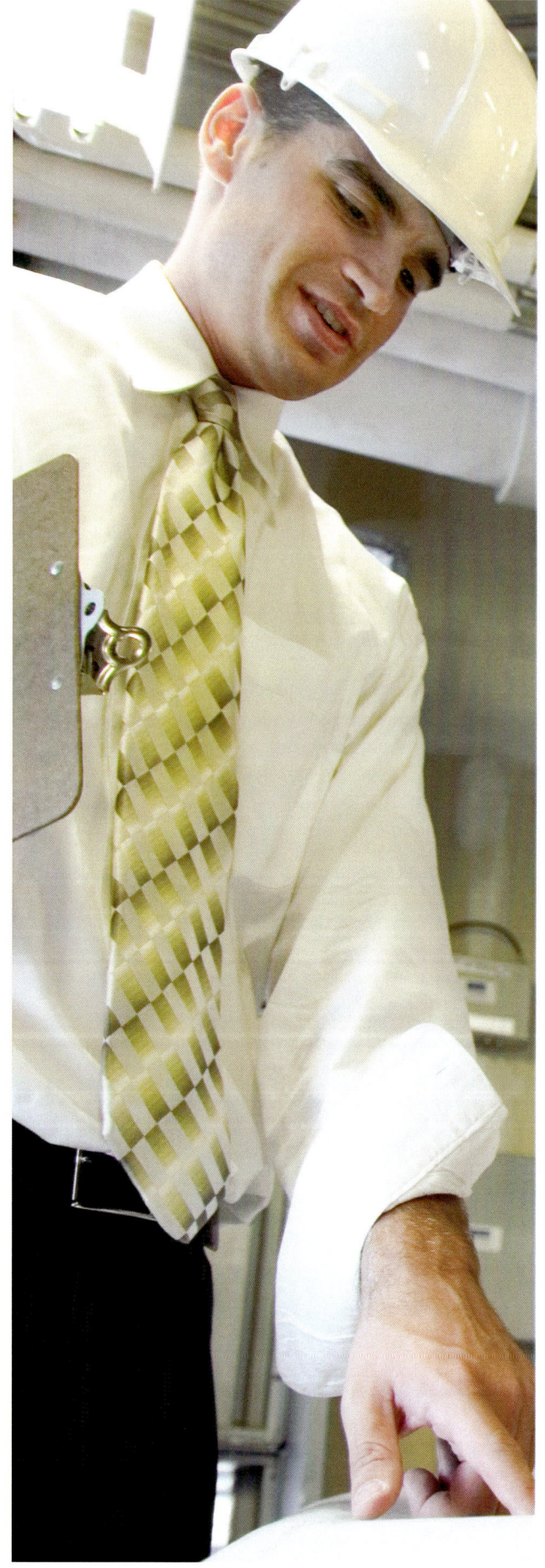

References

1 *Work with materials containing asbestos. Control of Asbestos Regulations 2006. Approved Code of Practice and guidance* L143 HSE Books 2006 ISBN 978 0 7176 6206 7
www.hse.gov.uk/pubns/books/l143.htm

2 *Managing health and safety in construction. Construction (Design and Management) Regulations 2007. Approved Code of Practice* L144 HSE Books 2007 ISBN 978 0 7176 6223 4 www.hse.gov.uk/pubns/books/l144.htm

3 *Managing asbestos in buildings: A brief guide* Leaflet INDG223(rev5) HSE Books 2012 (priced packs ISBN 978 0 7176 6487 0)
www.hse.gov.uk/pubns/indg223.htm

4 *Asbestos: The licensed contractors' guide* HSG247 HSE Books 2006 ISBN 978 0 7176 2874 2 www.hse.gov.uk/pubns/books/hsg247.htm

5 *The management of asbestos in non-domestic premises. Regulation 4 of the Control of Asbestos Regulations 2006. Approved Code of Practice and guidance* L127 (Second edition) HSE Books 2006 ISBN 978 0 7176 6209 8
www.hse.gov.uk/pubns/books/l127.htm

6 REACH (Registration, Evaluation, Authorisation & Restriction of Chemicals Regulations 2007) www.echa.europa.eu/web/guest/regulations;jsessionid=C2A247604D54B688F9735BCD1EB10436.live2

7 *Health and Safety at Work etc Act 1974 (c.37)* The Stationery Office 1974 ISBN 978 0 10 543774 1

8 *Management of health and safety at work. Management of Health and Safety at Work Regulations 1999. Approved Code of Practice and guidance* L21 (Second edition) HSE Books 2000
ISBN 978 0 7176 2488 1
www.hse.gov.uk/pubns/books/l21.htm

9 BS EN ISO/IEC 17020:2012 *Conformity assessment. Requirements for the operation of various types of bodies performing inspection* British Standards Institution

10 BS EN ISO/IEC 17024:2003 *Conformity Assessment. General requirements for bodies operating certification of persons* British Standards Institution

11 BS EN ISO 9001:2008 *Quality management systems. Requirements* British Standards Institution

12 BS 6002-4:2006 ISO 3951-5:2006 *Sampling procedures for inspection by variables. Sequential sampling plans indexed by acceptance quality limit (AQL) for inspection by variables (known standard deviation)* British Standards Institution

13 BS EN ISO/IEC 17025:2005 *General requirements for the competence of testing and calibration laboratories* British Standards Institution

14 *Asbestos in system buildings: Control of Asbestos Regulations 2006. Guidance for duty holders* HSE 2008 www.hse.gov.uk/services/education/claspguidance.pdf

15 *Asbestos: The analysts' guide for sampling, analysis and clearance procedures* HSG248 HSE Books 2005 ISBN 978 0 7176 2875 9
www.hse.gov.uk/pubns/books/hsg248.htm

16 BS EN 60335-1:2002+A15:2011 *Household and similar electrical appliances. Safety. General requirements* British Standards Institution

17 *Asbestos essentials: A task manual for building, maintenance and allied trades on non-licensed asbestos work* HSG210 (Third edition) HSE Books 2012 ISBN 978 0 7176 6503 7
www.hse.gov.uk/pubns/books/hsg210.htm

18 *Accreditation of bodies surveying for asbestos in premises* Edition 3 RG8 UKAS 2010 (for the application of ISO/IEC 17020)

Further information

For information about health and safety visit https://books.hse.gov.uk or http://www.hse.gov.uk. You can view HSE guidance online and order priced publications from the website. HSE priced publications are also available from bookshops.

To report inconsistencies or inaccuracies in this guidance email: commissioning@williamslea.com.

British Standards can be obtained in PDF or hard copy formats from BSI: http://shop.bsigroup.com or by contacting BSI Customer Services for hard copies only Tel: 0846 086 9001 email: cservices@bsigroup.com.

The Stationery Office publications are available from The Stationery Office, PO Box 29, Norwich NR3 1GN Tel: 0333 202 5070 Fax: 0333 202 5080. E-mail:customer.services@tso.co.uk Website: www.tso.co.uk. They are also available from bookshops.

Statutory Instruments can be viewed free of charge at www.legislation.gov.uk where you can also search for changes to legislation.